DOING BETTER STATISTICS IN HUMAN–COMPUTER INTERACTION

T0200692

Each chapter of this book covers specific topics in statistical analysis, such as robust alternatives to *t*-tests or how to develop a question-naire. They also address particular questions on these topics that are commonly asked by HCI researchers when planning or completing the analysis of their data. The book presents the current best practice in statistics, drawing on the state-of-the-art literature that is rarely presented in HCI. This is achieved by providing strong arguments that support good statistical analysis without relying on mathematical explanations. It additionally offers some philosophical underpinning for statistics, so that readers can see how statistics fit with experimental design and the fundamental goal of discovering new HCI knowledge.

PAUL CAIRNS is a Reader in Human–Computer Interaction at the University of York, UK, and Scholar-in-Residence for The AbleGamers Charity, which helps people with disabilities combat social isolation by making videogames more accessible. He has taught statistics at all levels of education for nearly twenty years. His particular research interest is in players' experiences of digital games, and his expertise in experimental and statistical methods was developed through working in this area.

DOING BETTER STATISTICS IN HUMAN–COMPUTER INTERACTION

PAUL CAIRNS

University of York

CAMBRIDGE
UNIVERSITY PRESS

CAMBRIDGE
UNIVERSITY PRESS

University Printing House, Cambridge CB2 8BS, United Kingdom

One Liberty Plaza, 20th Floor, New York, NY 10006, USA

477 Williamstown Road, Port Melbourne, VIC 3207, Australia

314–321, 3rd Floor, Plot 3, Splendor Forum, Jasola District Centre, New Delhi – 110025, India

79 Anson Road, #06–04/06, Singapore 079906

Cambridge University Press is part of the University of Cambridge.

It furthers the University's mission by disseminating knowledge in the pursuit of
education, learning, and research at the highest international levels of excellence.

www.cambridge.org
Information on this title: www.cambridge.org/9781108482523
DOI: 10.1017/9781108685139

First published 2019

Printed in the United Kingdom by TJ International Ltd. Padstow Cornwall

A catalogue record for this publication is available from the British Library.

ISBN 978-1-108-48252-3 Hardback
ISBN 978-1-108-71059-6 Paperback

For HB, my ground, my sky and my sunshine,
and for Samson, whom we miss.

Contents

Figures

Tables

Acknowledgements

This book is a culmination of a lot of thinking, teaching and thinking about teaching around statistics and accordingly there are lots of people who have had a hand in getting me to this point.

I would like to thank all of my PhD students, at York and at Goldsmiths, for their encouragement and enthusiasm for this project all the while I have been on sabbatical. I would like particularly to thank Joe Cutting for his concrete and constructive feedback on early versions of chapters. I would also like to thank all those research students explicitly mentioned in the book for giving me such rich, diverse and interesting research (and resulting data) to think about.

Many thanks go to my good colleagues Helen Petrie and Chris Power. Much of this book was written while on sabbatical and they took on my teaching load to make the sabbatical possible. Also, I would like to thank Chris further for his patience in listening to me enthuse about statistics and also forcing me to explain things properly. Thanks too to my department for allowing me to take the sabbatical.

Jan Böhnke has also been a helpful support over the years as I have grown my statistical knowledge. He has challenged my assumptions through many valuable discussions. Harold Thimbleby has been a great support throughout my academic career and his feedback on this book was the just right thing at just the right time.

I would like to thank my many students at all levels of higher education. They put up with me going on about statistics a lot but it is through my attempts, successful and unsuccessful, at trying to teach them statistics that I have gained a lot of the experiences that inform this book.

Finally, I would like to thank my family. Most of my family have been encouraging even in the face of what must have seemed like glacial progress in writing. Thank you for your support and belief. Two key members of my family who have been less supportive are my children, Eleanor and Patrick.

Their cheerful lack of interest in my work has been an important grounding both for this book and life generally. And last but most of all, thanks to HB for being my constant companion and friend and, also, for giving useful feedback on this book as it progressed.

Getting Started

Human–Computer Interaction (HCI) is concerned with understanding the relationship between people and interactive digital systems. It draws on diverse disciplines, in particular, psychology, computer science, social sciences and, increasingly, design, creative arts and critical theory. There are therefore occasional existential crises in HCI as to whether it is a science, a design discipline, a technical discipline, a social science or all of these or none of them (Reeves, 2015). Regardless of what HCI is or should be, it is clear that experiments, quantitative data and, therefore, statistics are regularly used to advance our understanding of how people experience, use and are used by digital systems.

However, many of us who set out to do HCI did not also set out to become statisticians. Typically, we have arrived at statistics, and the experimental methods that go with them, as the tools we need to tackle the research questions that interest us. Fortunately, there are many good textbooks and courses, particularly coming from psychology, that are well suited to supporting such HCI research, my personal favourites being Howell (2016) (currently in its 9th edition) and Pagano (2012). The problem is that textbook descriptions and examples are not really like the messy business of acquiring quantitative data and then trying to work out what to do with it. The fledging researcher very soon meets the reality of data that is not normal or has unexpected outliers and a plethora of tests that might (or might not) help. The Internet seems to help with useful comforting advice like t-tests are robust to deviations from assumptions and other such folklore. But then the Internet also suggests that left-handed haggis are dying out so you cannot always believe what you read on the Internet.

At this point, many researchers need a more experienced practitioner to talk through the specific questions they have about their data or the data they would like to collect. For many years, I have been teaching statistics and so I have ended up as just such an experienced person who people turn to. On the whole, the questions that people ask boil down to seeking reassurance and the general question of 'Am I doing this right?' Unfortunately, the answer is more often than not 'It depends'. Which of course begs the question of what it depends on. The purpose of this book is to provide a resource for researchers that hopefully addresses some of their questions about what might be the right way to gather and analyse data. What I am very keen to avoid, though, is simply providing a new source of folklore to compete with Internet 'wisdom'. For this reason, I have deliberately taken a quite scholarly approach of digging into the questions and trying to show where any existing folklore comes from, what the debate around current thinking is and, where possible, propose some possible answers or at least ways forward. What I rarely offer is definitive answers.

At the end of the day, when it comes to data collected with people and the uncertainty inherent in such data, it is very hard to have clear answers to statistical questions that provide the reassurance that people seek. The goal therefore is not to provide a definitive statement about what is the right way to analyse data but instead to help guide the reader to do a bit better and to be, at least, explicit about the compromises that have to be made.

Who Should Read This Book?

This book is intended for anyone who is about to collect or about to analyse quantitative data for an HCI study and is worried about whether they are doing it right. Understandably, in such data, you might have a whole host of questions and this book could not pretend to answer them all. Instead, I have tried to address the questions that I am asked most often or, in some cases, ones that I think HCI researchers should be asking more often.

As such, the reader is expected to be in a position to do HCI research but this could be anyone from a first-year undergraduate doing an evaluation to an experienced researcher meeting a new problem in their data. However, if you are undertaking a research project then I would hope you have read at least one book about planning quantitative studies, such as Harris (2008) and another about the standard statistical methods, such as Howell (2016). There are other excellent textbooks like these. You may also have had a course in experimental methods and statistics, which would be another very good place to start.

What I am not expecting is that you necessarily have, what might be called, advanced knowledge. On the whole, as discussed with reference to simplicity (Chapter 5), I do not think more sophisticated statistics are the solution to most (or possibly even any) of the problems that people ask me about. For this reason, the textbooks that I recommend are more or less accessible to anyone entering higher education. That is, your level of background knowledge should not be a bar to getting something out of any of the chapters in this book though you may have to do some homework to really get to grips with some parts.

Hopefully, right now, you will only be at the planning stage of your study. Some of the questions in this book, for instance about how many options a Likert item should have (Chapter 15), will help you to design your study well. In reality, I know that some of you will be reading this having designed an experiment, gathered the data and now are stuck trying to analyse it. I cannot promise that these essays will help you retrieve something useful from your hard work but they may help you navigate some of the most common problems such as the accusation of fishing (Chapter 13) or how to deal with outliers (Chapter 8). But don't do it again! And read the chapter about planning your analysis (Chapter 5).

This book could also be of use to researchers outside of HCI. However, the examples and practical concerns are from the context of my own research field and, in many cases, my own research experience. At the same time, the questions addressed could be relevant to any researcher considering the quantitative analysis of data arising from human participants. Thus, this book should be of use to you if you come from the psychological and behavioural sciences though some of the examples might not be so illuminating.

The Structure of the Book

As will become clear from this book, I do not see statistics as an end in themselves. Statistics are one part of a larger toolset that helps us to answer research questions. Statistics are inherently interwoven with the studies that generate the data to be analysed. The studies in turn are deeply connected with the new knowledge we seek. How knowledge, studies and statistics work together to form science is a matter of philosophy. While it might be simpler to say that such philosophy is irrelevant to practical statistics, I think one of the key things I am trying to say throughout this book is that you cannot really understand why statistics is done the way it is without thinking about why studies are done the way they are. For this reason, Part I of this book offers the philosophical backdrop to the more practical answers offered in Part II. I do not frequently encounter the questions that the chapters in Part I address but I do find that, regularly, the answers that I give about other questions need these more philosophical explanations. One of the best examples of this is the question 'Why can't I do all the statistical tests on my data that I like? After all, it does not change the data.' This seems reasonable but it flies in the face of how statistical tests work in concert with the experiment that generated the data being tested (Chapter 1). A knowledgeable person might then come back at me and say that that is not a problem for Bayesian statistics, it is just a flaw in how we traditionally do statistics. But that's not true! Unfortunately, not many explanations of Bayesian statistics actually address this issue properly (Chapter 3).

It is not essential that every reader read every chapter of Part I before tackling any chapter in Part II. In fact, you could probably safely launch straight into any of the chapters in Part II. But it just might be that to really answer the question you have, you will need to go back to the fundamentals in Part I of why we do statistical tests the way we do. Where I have anticipated this, I have indicated it in the chapters.

Each chapter in Part II is intended to stand on its own and is motivated by particular practical questions or issues relevant to gathering or analysing quantitative data in HCI. As such, each of these chapters addresses a set of questions around a topic, such as what do if data are not normal, and these questions are listed at the start of each chapter. If your question is listed then hopefully the chapter will help you to think about possible solutions to your problems. Ideally, you would not need to read several chapters at once: you should get some useful answers to the typical questions without reading further. In addition, each chapter is deliberately intended to be a short essay that targets the motivating questions so that even if you do not find the answers you seek, you will not have spent too long finding that out.

What you will not find are definitive answers. In some cases, there are answers that are best practice but even they are not necessarily the last word. I think the key to doing statistics well is to keep learning, to recognise the limitations of what we know and what we can do and to try to work out how we might do things a bit better. Definitive answers tend to give the impression that there is nothing more to learn so I do not try to give any.

Of course, in trying to answer some questions, we are led to further questions, and discussions of one issue may simultaneously address other issues. Thus, you will find in each chapter links to the topics and issues addressed in other chapters. Some chapters even thread together to make coherent units that would provide a broader background to more general topics in statistical methods. For example, some topic threads are:

1. Using modern robust statistics: Chapters 4, 5, 7, 10 and 11.
2. Problems with distributions or the shape of data: Chapters 7, 8 and 14.
3. Developing questionnaires: Chapters 15, 16 and 17.
4. Choosing suitable tests: Chapters 4, 5, 6 and 18.

I should emphasise that I chose the topics of the chapters to address the questions that I regularly get asked. As such, there is a degree of selectivity and possibly even idiosyncrasy in the topics of these essays. I do not aim, like the father of essays, Montaigne (1958), to be the 'matter of my book' but at the same time I recognise that the topics covered and the approach to those topics are based on my own particular experiences. For that reason, I do present my opinions but I will try to clearly indicate wherever it is my opinion in the absence of more authoritative viewpoints and also I am happy to write in the first person. These are my essays: my attempts to communicate statistics my way. The final test of the success of this approach is with you the reader. If you like this approach, let me know. If you do not, also let me know! And if you have other questions that this book does not answer, then I would be happy to hear from you and to see if I can help.

Statistics Software

Though occasionally it may be informative to perform the calculations of a statistical test with pencil and paper, in practice all statistical testing is done using statistics software in one form or another. Indeed, it should be this way: the most careful hand calculation can still have errors in it even with quite small datasets. Statistics software will not help you to do the right calculation but once you are doing the right calculation, the software will help you to keep doing it right.

Unfortunately, there is a plethora of software options for the practising researcher to use. These include:

- Well-established commercial packages such as SPSS, SAS and MiniTab
- Open source options like R and the statistics.py library for Python
- Function libraries as part of other data and mathematical software packages such as MatLab, Mathematica and Excel
- Bespoke online web apps, some of essentially unknown provenance and quality.

Because this book is not intended to be a textbook of statistics, I had hoped to side-step the issue of which package is best and in doing so avoid any ideological battles about what makes a good package. Part of my reasoning was also that it is a topic that I have found hard to care about. It seemed to me, as an HCI researcher, that your choice of statistical package boiled down to choosing the one that annoyed you least. They will all annoy you to some extent, through a diverse range of usability problems, but you will be able to tolerate some better than others depending on your particular experience and disposition. However, in writing this book and researching some of the most modern approaches, it is clear that R is becoming an important package that is starting to dominate statistical software.

There may be interesting cultural and disciplinary reasons for this. For instance, R is open source, which makes it easy for people to obtain and work with, unlike commercial software, which may not only require a licence but one that requires regular renewal. My own story is that originally I used SPSS because it was commonly used and available in the psychology department where I started teaching statistics. However, with each SPSS version that came out it seemed to get slower and slower and at the same time introduce more and more interface and interaction inconsistencies and problems. Even now I am still not sure what I will find under the SPSS menu option 'General Linear Model' as opposed to the 'Generalized Linear Model' and reliably finding a Mann–Whitney test is a thing of the past.

In recent years, I have therefore looked for other options and R has been useful, allowing a level of control and consistency that SPSS did not offer and at the same time ensuring that my students could do at home exactly what they had seen done in classes or labs. Moreover, it was free!

But these are not my reasons for recommending R. My main reason is that it became clear in my reading for this book that statisticians are using R as a way to embody their statistical knowledge. Whereas previously a statistician might outline a calculation or function in a paper and the reader would have to work out how to perform the calculation in practice, that outline is now backed up by R code, both as a listing and a downloadable package. The result is that you can apply it almost immediately if you are using R. A particularly important example of this is the WRS2 package developed by Mair and Wilcox (Available from: cran.r-project.org). This provides many of the statistical analysis functions that Wilcox promotes as best practice in modern statistics (Wilcox, 2017), including robust versions of the t-test and ANOVA that I discuss in Chapters 11 and 12.

This is not to say R is perfect. Far from it. Its language is a computer scientist's nightmare, being a bastardisation of list-processing, weakly-typed, object-oriented and procedural paradigms with unflagged case-sensitivity thrown in for good measure. As I said, all statistics packages are annoying. Nonetheless, it is the environment where cutting-edge, best-practice statistics are being made available.

I would therefore strongly recommend that you get familiar with R so that you are in a position to use the best techniques and methods, not only those that are currently out there but those that may be developed in future. I tend to use R through the RStudio interface because that just helps to organise scripts, output and workspace. There are good online resources and textbooks to help the new user. I've found Quick R (www.statmethods.net) useful because that assumes you have good experience of statistics and just want to transfer that over to R. I have also found Matloff (2011) useful to help make sense of the almost arbitrary semantics of the R language. And if you are coming to R and statistics afresh then I can also recommend the R version of Andy Field's book (Field, Miles, and Field, 2012) and if you like your statistics textbooks to have a healthy dose of steampunk you might even like Field (2016). Learning R may not be straightforward but it is worth the effort.

References

Field, Andy (2016). *An Adventure in Statistics: The Reality Enigma*. Sage.

Field, Andy, Jeremy Miles and Zoe Field (2012). *Discovering Statistics Using R*. Sage.

Harris, Peter (2008). *Designing and Reporting Experiments in Psychology*. McGraw-Hill Education.

Howell, David C. (2016). *Fundamental Statistics for the Behavioral Sciences*. Nelson Education.

Matloff, Norman (2011). *The Art of R Programming: A Tour of Statistical Software Design*. No Starch Press.

Montaigne, Michel de (1958). *Complete Essays of Montaigne*; Trans. by Donald M. Frame. Stanford University Press.

Pagano, Robert R. (2012). *Understanding Statistics in the Behavioral Sciences*. Cengage Learning.

Reeves, Stuart (2015). 'Human-computer interaction as science'. *Proceedings of the Fifth Decennial Aarhus Conference on Critical Alternatives*. Aarhus University Press, pp. 73–84.

Wilcox, Rand R. (2017). *Introduction to Robust Estimation and Hypothesis Testing*. 4th edn. Academic Press.

PART I

Why We Use Statistics

CHAPTER I

How Statistics Support Science

▷ Do I need a theory before I do an experiment?
▷ How can my experiment prove anything?
▷ How can statistics prove anything?
▷ Don't I need huge amounts of data to be sure of anything?

The purpose of this book is to answer questions about how to do statistics better. Doing statistics well, though, is not an end in itself; rather it is about engaging in a process that allows us to find stuff out. Sometimes this process is grandly labelled as doing SCIENCE (pronounced just like science but in a resonant, deep tone and with an obvious capital letter) but I prefer to think of it without the official label. There are lots of ways of finding stuff out and science is one, admittedly quite effective, way. Statistical reasoning is a useful tool for HCI because we often have to work with data that are not necessarily very precisely defined or measured, being about people's subjective experiences, or else they are subject to random influences of context that we cannot control. Using statistics, we are able to peer through the inherent uncertainty in our data to begin to see how things work.

In SCIENCE, the stuff we are trying to find out is often grandly formulated as a theory, a description of how the world (or part of it) works and with which we can predict how it might work in future. This is something of a challenge for HCI because we do not really have a lot of theories of interaction: it is all just very messy and complicated. There are notable exceptions, such as Fitts's Law and Information Foraging theory, as others have discussed (Reeves, 2015). On the whole though, HCI does not have substantial bodies of theory but nonetheless there are lots of interesting things still to find out: why do people make errors with interfaces? what aspects of interface make it more aesthetically pleasing? what is the best way to provide feedback to users? Statistics can help us to do that by allowing us to interpret our data clearly. Even so, this process of moving from data to new knowledge is tricky and one that has been difficult to sort out even in SCIENCE. It is called the problem of induction.[1]

In this chapter, I would like to discuss the problem of induction as seen in the philosophy of science. This might look like an attempt to demonstrate why HCI is a science but first, that would be a much larger endeavour and secondly, I am not sure how useful it would really be. Instead, the goal here is to discuss one solution to the problem of induction called severe testing (Mayo, 1996). Statistical analysis fits well into the framework of severe testing and so this provides a useful philosophical context for understanding how good statistics help to address the problem of induction.

Alongside severe testing, this chapter will also introduce the philosophy of science movement called new experimentalism. This viewpoint regards experiments as first-class results of scientific activity, to stand alongside theory in the framework of science. This is useful to us as it helps to understand and, at least, value the strong emphasis in HCI on empirical work despite the absence of substantial theories (Hacking, 1983). First, though, we need to see what exactly is the problem of induction.

1.1 The Problem of Induction

Put simply, the principle of induction is: having observed lots of examples of something, there is a law that says that is what always happens. The classic example is that every day in my life, I have seen that the sun comes up[2] and therefore I conclude that there is a law (of nature) that every day that sun will come up (Chalmers, 1999). The problem of induction is: what is the logical basis for this law? That is, how do we induce (as opposed to deduce) the behaviour of the world in general from the specific behaviour that has been observed? Mathematically, the problem of induction is solved by building an inductive principle into the fundamental axioms of mathematics (Halmos, 1968). In the natural world though, there is no such logical link. We have theories but how are those theories supported by the finite set of observations, facts or data that we have?

---

[1] The problem of induction is a problem for all knowledge, not just science, but it can be particularly problematic for science where its theories purport to have universal and general applicability.
[2] Living in rainy England that's not strictly true or even approaching true but let's not worry too much about that…

We may have very good theories about why the sun rises. Ancients may have thought it was because of a powerful god following a heavenly prescribed path. Others have had theories of celestial spheres revolving in harmonious perfection around the Earth. We currently have the theory of a rotating Earth alongside theories of gravity. Whatever the theory, none of them logically entail the sun rising tomorrow, because all such theories cannot speak to what else might happen. A new, more powerful god may rise to smite the sun, the celestial sphere might shatter or the solar system may move into a part of universe where gravity behaves differently, say a particularly thick patch of dark matter. We just do not know.

Nonetheless, we tend to behave as if our theories hold reliably and indeed such behaviour is warranted. We plant seeds, turn keys in cars, get on aeroplanes and shop online in the very reasonable belief that the underlying physics, chemistry, biology, engineering and so on will all work to produce the outcome we expect.

However, when it comes to producing a new theory where we are not surrounded by repeatable, robust phenomena, inducing (as opposed to deducing) the nature of the theory from the data becomes very challenging. There are literally millions of theories that can produce the same data, regardless of how much data we have. Which out of all of them is right? It is clear that there are precisely correct but generally wrong theories such as the ones that define precisely the data you get in a particular experiment but without relevance to any other experiment. And there are also very vague theories ('It rains at least once a month', which is definitely true in my life) that are not particularly useful because they are so vague. And it is clear that setting up situations that confirm the theory add little weight because if we already have data supporting a theory, more such data does not tell us more about the theory.

It was against this problem that Karl Popper proposed the framework of falsification (Popper, 2005) wherein no theory could be proven but only falsified by data that did not agree with it. Failing to falsify did not prove a theory but perhaps suggested that the theory had some robustness. This has two practical consequences for finding things out. First, a useful idea needs to have sufficient detail to be demonstrably wrong, that is, it is in principle falsifiable. Astrological predictions often fail to be falsifiable because they are not specific enough to be obviously wrong. Second, if you want to 'prove' a theory then you do not look where theory has been shown to work but deliberately seek out opportunities for it to fail. The logical problem with falsification is that no theory lives in isolation but sits on a stack of underpinning theories including those needed to explain any devices that gathered the falsifying data, including our own senses. So when a datum falsifies a theory, exactly which part of the theory is it falsifying? (Chalmers, 1999). In particular, historically, it is seen in science that even in the face of apparently falsifying evidence, a theory is still defended and it is the evidence that is questioned. Popper's framework does not explain how to resolve such arguments in any systematic way.

Around the same time as Popper, Kuhn showed how science could be seen as a socially constructed endeavour (Kuhn, 1975) where scientists collect their effort around a socially accepted paradigm in which to work. Normal science, that is science that fits within the currently accepted paradigm, works to explain 'puzzles' that are not yet currently explained by the paradigm. Solving the puzzles reinforces the paradigm and failure to solve them could be attributed to a failing of the scientists rather than the paradigm. Anomalies that defy explanation are possible but only in so much as it is expected that, within the paradigm, they will eventually be accounted for.

However, over time there could be an accumulation of new data or persistent anomalies that remain puzzles and resist solution within the paradigm. When there are enough anomalies, the existing paradigm falls apart and this leads to the now famous 'paradigm shifts' in which there would be a revolution from one paradigm to another. A change of paradigm completely changes both what is normal and what are the legitimate topics to study. In the new paradigm, what was once considered normal and mainstream science becomes sidelined and irrelevant. There are definitely powerful insights in Kuhn's work, in particular that scientific work tends to be conducted within a socially accepted frame (this is actually a problem for statistical analysis, as discussed more in Chapter 2). However, it is very unclear how the paradigm shifts work. Indeed, it seems that even the exemplar paradigm shifts that Kuhn identified, such as the shift from the phlogiston to the oxygen paradigm of combustion, do not really bear up to the historical facts of what happened (Chang, 2012).

Thus, if there is no logical or even systematic way of proceeding from data to knowledge, how then does science, or scientific knowledge more broadly, manage to progress at all? Feyerabend even went so far as to say there was no such thing as scientific method, just the tastes and freedoms of individual scientists (Feyerabend, 1993). This clearly sits rather uncomfortably with the grand vision of SCIENCE as a special discipline. This is not necessarily a problem, but it also feels wrong when thinking about the more day-to-day practice of scientists. Human endeavour has progressed substantially over the millenia and things like farming, the Internet and space travel speak to a growth in both the extent of our collective knowledge and the sophistication of that knowledge. It seems too much to think that this is just due to the chance wanderings of personal inclinations.

1.2 Severe Testing

One approach to the systematic growth in scientific knowledge is the notion of severe testing put forward by Mayo (1996). In this framework, there are theories or ideas that people put forward. Why these ideas are put forward is irrelevant. Theories and ideas might grow from existing data and facts or be inspired in some way from them but that is not necessary to a theory. Some theories and ideas might be little more than the hunch of an individual researcher. What gives ideas merit is that they can be put under a severe test and shown to withstand that test. The exact nature of what constitutes a test and how severe it is negotiable within the community of researchers.

At first glance, this may seem like falsification again but it is not. In the severe-testing framework, only the idea under scrutiny is severely tested. The processes, devices and measurements used to test it are not severely tested: they are expected to work in some systematic way in order to construct the test. If an idea fails a severe test then that counts as evidence against the idea. However, it can be argued that it counts as evidence against the testing procedure in which case specific severe tests of the test itself need to be conducted. A really good test is one where the ideas that underpin the test are already firmly established (have been severely tested) and so likely to resist any further severe tests.

This can be made concrete with an example drawn from my own research. There is a theory of aggression in digital games that is used to account for how violent video games lead to aggressive behaviours. This is called the General Aggression Model or GAM (Anderson and Bushman, 2002). The GAM, as might be expected, is quite an extensive model that moves from the act of playing a game to the actual aggressive behaviours of people. It would be challenging to severely test all of the ideas embodied in the GAM in one go. Therefore, my then PhD student, David Zendle, worked on severely testing just the first step of the GAM, which says that as a result of playing violent video games, players are primed with the violent concepts. Priming is a psychological phenomenon where, when people are exposed to one concept, such as guns, they are quicker to respond to related concepts such as death, killing, soldiers and so on (R. J. Sternberg and K. Sternberg, 2016).

The GAM predicts that increasing the realism of the depiction of violence increases the priming for violent concepts. In one of David's experiments (Zendle, Kudenko and Cairns, 2018), players of a game are exposed to two versions of the game, one that depicts more realistic behaviour when opponents are shot (using ragdoll physics), and one that displays standard animations when opponents are shot. Players are then measured for the extent to which they are primed for aggressive concepts using a standard measure of aggressive priming, namely a word-completion task (Anderson, Carnagey, and Eubanks, 2003).

This experiment acts as a severe test of the GAM in various ways. It takes a concrete prediction of the GAM, that players are primed for violent concepts by the realism of the violent content and examines that. Of course, we do not know to what extent people normally have access to violent concepts, so those exposed to the more violent depiction are compared to those who do not see the more violent depiction. Also, we know that even playing a game could increase arousal and arousal influences priming, so the two groups of people play essentially the same game except for this one feature. We even take care to change only the animation of when an opponent is shot so that it will not otherwise interfere with the gameplay. Thus, by making a careful comparison between two groups of people, the experiment becomes a severe test of one small part of the GAM. And by using standard games and standard measures, we are assuming that these have established validity as instruments in the experiment and are not in question. Thus, the only idea under test is the idea we are interested in.

And the result? We actually found the opposite of the GAM – that if anything, players were suppressing violent concepts as a result of playing the game with more violent depiction rather than being primed. Thus, the GAM failed to pass this severe test. Of course, this does not refute the GAM. Others might argue that our test was invalid in some way that we have not yet noticed, though obviously we would try defend against this. Alternatively, this particular aspect of violence is not core to the GAM. Either way, the experiment provides evidence that the GAM is not sufficiently nuanced. More experiments, more severe tests, should clarify in what way it needs to be improved or even if it needs to be substantially changed.

From this description of severe tests, it should be clear that experiments are essentially a form of severe test, a way of exposing ideas so that if they are right they are clearly right and if wrong they are clearly wrong. It is interesting that for decades (or more) researchers of all sorts have known how to set up experiments as severe tests but without calling them such.

Severe tests do not have to be experiments. To think about these, I always refer to astronomy as it clearly is not an experimental science although it is a science. Severe tests in astronomy consist of deliberately making observations to see things where theory, hunches or ideas say they should be. For example, the theory of planetary formation in our own solar system suggests that planets should form regularly around stars like our own sun but planets, not being bright like stars and also considerably smaller, are hard to see. This led astronomers to devise what would be signatures of exoplanets in the light coming from other stars. And now finding exoplanets is a citizen science project that everyone can join in.[3]

Thinking about the structure of severe tests, it should also be clear that you cannot simultaneously test an idea and the method used to test it. This is something done surprisingly often in HCI, where the experiments used to demonstrate (severely test) an idea are also the ones used to establish the method of measurement.

1.3   Evidence in HCI

Having put some philosophy of science in place, it is useful to think how this fits in the context of HCI, particularly the types of HCI where we use statistics. This is not intended to be the last word on the matter and references back to this philosophy will be made wherever necessary in the other chapters in this book. Here I will use statistics informally to mean a bunch of calculations and methods that are able to identify patterns in uncertain data that are unlikely to be due chance alone. I will talk about exactly what that means in other chapters, most particularly in Chapter 2.

The severe testing framework highlights the importance of having an idea to test but the idea need not be so elevated as to be dubbed a theory. This is particularly useful for HCI because often we have very concrete things that we need to understand but their context and use is so complex that there are no universal theories we can call on to explain them (Harrison, Sengers, and Tatar, 2011). Nonetheless, with severe testing we are in a position to grow our knowledge even in the absence of a strong theory.

Having clearly specified the idea to test, the test itself must also be specified and this must be tightly tied to the idea under investigation. However, often in the discussion of statistical analysis in textbooks, the data that are going to be analysed are assumed to have been gathered already

---

[3] www.planethunters.org.

through some lightly sketched process. This is understandable where the focus is on the mechanics of doing statistics but, in reality, most datasets would not come into being were it not for some experiment that had been *specifically* designed to generate them. Good statistics are intimately tied to the experiments that generate them and the two together form the basis for an evidential argument (Abelson, 2012). Experiments in HCI rarely generate self-evident data so statistics are needed to interpret them. At the same time, statistics are meaningless without the process that generated the data. In particular, if the experiment was set up to test one idea, it is not very useful to use these same statistics to test a different idea because the other idea is unlikely to have been severely tested by the data-gathering process. For example, if David Zendle had found in his aggressive priming experiments that players had taken more shots in the more realistic version of the game, this is not a test that more realistic graphics lead to more aggressive in-game actions because this idea simply had not been predicted and, if it had, an entirely different experiment might have been produced that did more manipulations of graphical realism than the ones used in his study.

Severe testing also suggests that the most severe tests are conducted by doing experiments. That is, the best severe tests explicitly intervene to manipulate the world and see how an idea performs in the resulting situations. Though broadly true, this does not preclude observational or survey studies from providing good knowledge through severe tests, just as observations underpin astronomy. For example, in HCI we may have very good reasons to think that personality influences people's ability to use a particular interface. We cannot influence personality (at least not ethically) but based on our ideas we could seek situations where this influence could be seen. Thus, a survey or observational study can severely test an idea when the study has set out to specifically gather data on that idea. Statistics can show all sorts of things about survey data, some of which will definitely be a result of chance variation, but it is clear that where statistics are used to specifically address the idea under test then they provide a greater evidential basis than a chance, unanticipated finding.

1.4   New Experimentalism in HCI

In the early accounts of scientific method, theory played a dominant role. Theories, however arrived at, are set up to be confirmed, falsified or severely tested. However, in more recent accounts of science, there has been a growing recognition of the importance of experiments in themselves as primary constituents of science. I think that the best example of this from the history of science is Faraday's motor. At the time at which Faraday developed his motor, there was only some rather informal accounts of the link between electricity, magnetism and movement. What Faraday did, with a very simple set-up, was to demonstrate this link in an incontrovertible way with a wire spinning around a magnetic so long as there was electricity running through it. It is an elegantly simple demonstration and there are some nice videos of modern versions on YouTube.[4] What is important for us though is that at that time Faraday had no theory of electromagnetism. In fact, his subsequent theory based on magnetic lines of flux was not right either. However, the absence of theory is no bar to the importance of Faraday's motor because, whatever theory of electricity and magnetism you come up with, it has to account for the motor.

What experiments like this do is isolate a phenomenon (Hacking, 1983). Such phenomena can stand as a severe test of a theory even before any theory is proposed. I call experiments that sharply define new phenomena first-class because they stand at the first level of scientific objects just like theories rather than derived or subordinate to theories.

Not every experiment has such a first-class status. David Zendle's experiment for instance works well as a severe test of the GAM but sits very much in that context. It does not provide new evidence of how aggression might arise from games, it only undermines one account of that process. This is not to say it is not good work, just that its reach is limited to testing the specific theory under scrutiny. It perhaps offers some evidence of how priming works in games but, if that really had been his focus, the experiment would probably have looked somewhat different.

Within HCI though, I think there are experiments that isolate phenomena that must be taken into account. One example is from the work done at UCL into the errors people make with interactions (Ament et al., 2013). When a person is using an interface to achieve a task, of course, they might make errors. However, some errors might be due to the interface, some to the task and some to a combination of both. In any interaction, it is very difficult to disentangle what might be due to the interface and what to the task. Training should help but it does not help reliably. And only the most coarse and unsubtle feedback works to reduce errors. Why is this? What Ament et al. (2013) showed was that you can give people two tasks that in terms of the interaction are *identical* but in which people make errors differently depending on how the task is framed.

---

[4] For example www.youtube.com/watch?v=MRFqYRHT3Wk.

I think this is a lovely example of isolating a phenomenon in HCI. No matter what account you have of errors, how people store tasks in memory, how feedback on interfaces works, how interfaces are laid out and so on, this experiment shows that what people think the task is alters the errors they make. It is not at all just to do with the design of the interface. So any future researchers looking into error with interactive systems have to account for this phenomenon and think just as carefully about the task design as well as the interface design. Of course, being good researchers, Ament and her colleagues offer their own account of what is going on and it may well be right but even if not, that would not undermine the experimental findings.

Even in this work, statistics were needed to see whether the differences observed were different enough to be judged as systemic in some way. Some people made more errors than others and errors were found in both conditions of the experiment so the picture of error behaviours was complicated. Severe testing allows us simply to have the idea that the task structure matters but it is the statistics that help us to see whether the idea passes the severe test. In some sense, relying on statistics weakens the first-class nature of the experiment as it weakens the clarity with which it isolates the phenomenon. But at the same time, in the face of natural variation between participants, this is just something we have to live with.

This more generally does challenge whether experiments that require statistics can provide first-class examples of isolated phenomena. Any such phenomenon that is seen as a result of statistics could just be an artefact of the data. But experiments that repeatedly show the same artefact with different people and in different settings can elevate an interesting artefact to an important phenomenon. The lesson is that we should not rely on one experiment, however compelling it may seem.

As well as doing experiments, HCI researchers often build new interactive systems: new types of display, new types of input, new types of interactive styles and so on. Sometimes these act as designs that show what is possible but sometimes they function as first-class objects that demonstrate something not previously seen. For example, W. Thimbleby and H. Thimbleby (2005) produced a completely novel touch interface for calculators. Though it was designed in a principled way and is used to articulate further principles of interaction, there is not necessarily any theory in a formal sense that says why these principles are sound. However, if these principles were explored and severely tested then the novel calculator would stand as an object that presents specific phenomena that should also withstand the severe testing. It may be grandiose to claim the status of SCIENCE for such designs but at the same time if people were to theorise around how designs like this work, then the designs present evidence that needs to be accounted for. Building interesting things produces evidence of phenomena and a basis on which a science of HCI can grow.

1.5 Big Data

It is worth mentioning here how statistics can fit with Big Data as this is something that is very current in many disciplines. Big Data has a multitude of meanings but is often characterised by some Vs, originally three Vs (Laney, 2001):

- Volume: lots of data
- Velocity: arriving all the time
- Variety: in lots of different, possibly incompatible forms

There are other characteristics proposed as well but these suffice here. HCI is becoming awash with Big Data from the details of purchases online, through the social networks people use, to the patterns of playing and spending on mobile games. Modern online technology offers unprecedented insights into what people get up to when they use interactive systems. However, it should be noted that the data usually stops there. While we may know very precisely what people are doing, Big Data rarely includes the information as to why they are doing it.

The problem for statistics, of any sort whether sophisticated machine learning algorithms or traditional t-tests, is that with the very large, complex and potentially idiosyncratic datasets that exemplify Big Data, there will always be artefacts that look unlikely to have occurred by chance. The problem always arises is if you look hard enough and are agnostic about what might be interesting: one in a million chances happen more often than you think (Pratchett, 1987, p. 7).

Severe testing helps us see the scientific value of Big Data more clearly. If you have found an unexpected outcome in your big data, having looked for any unexpected outcome, you have not really severely tested that outcome. For example, suppose the outcome is that people make no errors with your user interface on Tuesdays. If you would be happy with any day of the week or reduced task time or double the average number of errors, then this finding provides no evidence for a systematic effect. It could just be one of those things.

Of course, if you had expected reduced errors on Tuesdays, for instance because the user group on Tuesdays is restricted to people with specific skills, then obviously this result provides evidence in support of your idea *because you specified it without knowledge of the data*. Even then, you would need statistics to say that this was unlikely compared to error rates on other days. Notice that there is not a temporal component to this: the data could have been gathered before you formulated your idea. You went seeking a situation where your idea would be exposed to a test but without using the data to say what you should be looking for.

If however you found this result and then formulated your idea, then the data do not provide evidence, because you only had the idea on the basis of the evidence provided by the data. Seeing that the data fit your ideas does not test your idea in anyway, it just says you have suggested something that fits. At most, you can only remark on the interesting phenomenon in your data and propose your idea speculatively. To test the idea, you would have to seek new evidence.

From an HCI perspective, we are rarely interested in properties of our data. They lack intrinsic value to us. We are more interested in the experiences of the people using a system and how the characteristics of the system influence the user. While Big Data can help us to identify possibly interesting things, as HCI researchers, we are probably going to step back from Big Data and work out why people are behaving in this way. No amount of statistics can tell us what people are thinking and experiencing unless that is already inherent in the data.

1.6   Conclusions

When reasoning in the face of uncertainty, as is usual for statistics, we need to have some basis for moving towards increasing certainty in our ideas. This is the problem of induction and it can never be wholly removed. Severe testing provides a framework for making that move from having an idea to having evidence for the idea, whether statistically driven or not. Severe testing puts ideas into, or seeks them in, situations where if they were wrong they could be clearly seen to be wrong or if right clearly seen to be right. Passing a severe test provides evidence for the idea, but multiple tests are needed to fill out the scope and reliability of the idea. There is no ultimate proof in science ever, just ideas that keep passing severe tests.

Sometimes, our ideas are far from firm but nonetheless they form the basis for interesting experiments and interesting designs. These things can isolate phenomena that were not previously apparent. They may not offer explanations but they represent first-class objects that require explanation.

Philosophy of science can feel a long way from the actual practice of research. After all, scientists have been successfully getting their hands dirty and growing knowledge without having any good accounts of how what they are doing works. This is encouraging: scientists must have been doing something right for their ideas to flower. It is useful to have these accounts to fall back on when negotiating tricky problems in our reasoning, but now we need to get on and cultivate our own garden.

References

Abelson, Robert P. (2012). *Statistics as Principled Argument*. Psychology Press.

Ament, Maartje G. A. et al. (2013). 'Making a task difficult: evidence that device-oriented steps are effortful and error-prone'. *Journal of Experimental Psychology: Applied* 19.3, pp. 195–204.

Anderson, Craig A. and Brad J. Bushman (2002). 'Human aggression'. *Annual Review of Psychology* 53.1, pp. 27–51.

Anderson, Craig A., Nicholas L. Carnagey and Janie Eubanks (2003). 'Exposure to violent media: the effects of songs with violent lyrics on aggressive thoughts and feelings'. *Journal of Personality and Social Psychology* 84.5, pp. 960–971.

Chalmers, A. F. (1999). *What Is This Thing Called Science?* 3rd edn. Open University Press.

Chang, Hasok (2012). *Is Water H2O?: Evidence, Realism and Pluralism*. Vol. 293. Springer Science & Business Media.

Feyerabend, Paul (1993). *Against Method*. 3rd edn. Verso.

Hacking, Ian (1983). *Representing and Intervening: Introductory Topics in the Philosophy of Natural Science*. Cambridge University Press.

Halmos, Paul (1968). *Naive Set Theory*. Springer.

Harrison, Steve, Phoebe Sengers and Deborah Tatar (2011). 'Making epistemological trouble: third-paradigm HCI as successor science'. *Interacting with Computers* 23.5, pp. 385–392.

Kuhn, Thomas S. (1975). *The Structure of Scientific Revolutions*. 2nd edn. University of Chicago Press.

Laney, Doug (2001). '3D data management: controlling data volume, velocity and variety'. *META Group Research Note* 6, p. 70.

Mayo, Deborah G. (1996). *Error and the Growth of Experimental Knowledge*. University of Chicago Press.

Popper, Karl (2005). *The Logic of Scientific Discovery*. Routledge.

Pratchett, Terry (1987). *Mort*. Corgi Books.

Reeves, Stuart (2015). 'Human-computer interaction as science'. *Proceedings of the Fifth Decennial Aarhus Conference on Critical Alternatives*. Aarhus University Press, pp. 73–84.

Sternberg, Robert J. and Karin Sternberg (2016). *Cognitive Psychology*. Nelson Education.

Thimbleby, Will and Harold Thimbleby (2005). 'A novel gesture-based calculator and its design principles'. *Proceedings 19th BCS HCI Conference*. Vol. 2. BCS, pp. 27–32.

Zendle, David, Daniel Kudenko and Paul Cairns (2018). 'Behavioural realism and the activation of aggressive concepts in violent video games'. *Entertainment Computing* 24, pp. 21–29.

CHAPTER 2

Testing the Null

Questions I am asked:

▷ What are *p*-values really?
▷ Isn't it now thought that significance testing is really bad?
▷ Why is the null hypothesis so important when all it says is that nothing happens?
▷ My result isn't signficant: what should I do?

The mainstay but also the black sheep of statistics is null hypothesis significance testing (NHST). This is the classic process familiar to, if not always understood by, anyone who has ever looked at statistical analysis. A complicated dataset is reduced to a single *p*-value, a probability, usually produced via one or other well-established statistical test. The interpretation then is that small *p*-values, where small typically means less than 0.05, are declared to be significant. Significant results are good, non-significant results are bad.

NHST is the mainstay of statistics in that it really is how a lot of statistical analysis is done in a wide-range of research fields. HCI, drawing on its roots in psychology, has done a lot of experimental work in the NHST tradition and any experimental paper in HCI will be awash with *p*-values and the word 'significant'. This includes my own work.

The long-recognised problem of NHST though is both that it encourages incautious reasoning (Cohen, 1994) and, worse, leads to unthinking adherence to the process (Gigerenzer, 2004). Of course, any analysis method when practised ritually rather than with a view to careful interpretation can be a problem for science. However, the emphasis on *p*-values has bigger problems because of the social construction of the importance of *p*-values for analysis. There is an expectation amongst researchers and reviewers that the right way to do statistics is to seek small p-values. Teachers teach the importance of p-values, researchers produce p-values and reviewers expect to see them. In fact, researchers who do not provide them risk being rejected (Gigerenzer, 2004). This leads to the importance of reaching that crucial point where $p < 0.05$ and declaring signficance. The obsession with reaching this value has led to what has become known as p-hacking (Simonsohn, Nelson, and Simmons, 2014): making sure that the p-value falls below the all important 0.05.

It is not necessarily the case that people who have engaged in p-hacking are malicious or unscrupulous. It is just that modern academia places tremendous pressure on researchers to publish no-matter-what. Combine this with the socially constructed importance of finding $p < 0.05$ and people will consciously, or more often unconsciously, move to achieve this by analysing their data until that significance appears. Moreover, equally problematic is that if the results are persistently non-significant then the researchers will not even attempt to publish them. This is the file-drawer effect, where good quality experiments that might provide interesting severe tests of ideas (Chapter 1) are not published because of a lack of significance (Rosenthal, 1979). Work that leads to p-hacking is more likely to be a problem of people who do not really understand what statistics are doing in the first place rather than researchers deliberately trying to deceive.

Despite these obvious sociological problems of NHST, I am going to discuss why significance testing can be a valid approach to the analysis of data in HCI. I am not advocating the application of NHST as a mindless ritual, far from it. There is definitely a need for researchers using NHST to 'up their game' to use effect sizes, think about outliers, use more appropriate tests and, most of all, to reason carefully about their findings. But I think an improved NHST is good enough for us to make progress. It just requires, like all science, honesty in the researcher.

2.1   The Basics of NHST

Aside from the social problems of a fixation on significance, there are well-understood problems of what exactly p-values mean, and these confusions are as much a property of teachers as students (Haller and Krauss, 2002). The consequences of not understanding p-values are well recognised and even illustrated in the context of HCI research (Robertson and Kaptein, 2016a,b). So before really working out how to make NHST work, it is worth being clear what we mean by NHST. What follows is a breakdown of the process of testing a null hypothesis, which is implicit in many statistical analyses but which is also described in good textbooks, for example Howell (2016).

First, I would like to emphasise that NHST is a way of arguing from data to provide evidence where there is uncertainty in the data. That uncertainty may be due to imprecise measurements or to the unknown influence of context. These are concerns familiar to anyone who has tried to measure user experience. NHST does not, indeed cannot, prove or demonstrate anything but it helps the researcher justify a conclusion from the data and so grow the evidence for an idea.

For good NHST, you have to be clear about what the idea is you are trying to test. This is often called the alternative hypothesis, being the alternative to the null hypothesis, which I will come to in a moment.

Once, the alternative hypothesis is clear, the next step is to gather data. We would expect, if our ideas are right, that the data will offer support for the alternative, for instance by being predicted to look a certain way if the alternative hypothesis is true. Often, and for the purposes of our discussion, the data will come from an experiment set up explicitly to test the hypothesis. To be concrete, let's suppose we believe time pressure is important in the experience of immersion when playing digital games. This means we have to devise an experiment to gather immersion data under the influence of time pressure. The experiment will have all participants play a game but there are two conditions, one in which players have very limited time to reach a goal and one where the time limit is more generous. These are the experimental conditions and time pressure is the independent variable that is varied between the conditions. One condition has low time pressure, the other condition has high time pressure. Immersion is measured using a questionnaire and is called the dependent variable. The statistical alternative hypothesis is that increased time pressure will lead to increased immersion.

Typically, we would not expect every participant to achieve the same level of immersion in each condition nor that in one condition everyone's immersion would be higher than in the other. Instead, we need to think statistically. Each participant will produce a different immersion score depending both on their own experiences of immersion and their perceptions of that experience in relation to the items of the questionnaire. It is not the individual scores that matter but that, on average, the players have higher immersion scores in the high time-pressure condition than the other. Again, without getting too bogged down in experimental details, we can take 'on average' to imply the mean immersion scores.

The problem is that even at the level of means, we would still expect to see differences between the conditions whether or not our alternative hypothesis was correct. We would be very surprised if both conditions had exactly the same mean immersion score. In fact, I have spotted mistakes in my own papers because two means were identical. Fortunately, I did so before the papers were sent out for review.

Having gathered the data that test the alternative hypothesis, we set up a null hypothesis. This is the statement that there is no effect of the independent variable on the dependent variable. In our case, this would be that the mean immersion scores are not very different. 'Not very different' is a beguilingly vague phrase. We know we do not expect the means to be identical but how different is not very different? To make this precise, we also need to assume how our data should be distributed under the null, or more precisely how our means should be distributed. The term distribution is hiding some sophisticated mathematics, probability theory, about how to assign probabilities to random variables taking on particular values. Though very rarely stated in experimental work, the assumption of how data (or the statistics based on the data) are distributed is crucial to good analysis.

Distributions

The distribution of data is really a theoretical concept that is usually an assumption underlying any statistical test. In informal terms, the distribution is the probability that particular values of data are likely to turn up. To make this more concrete, let's start with one of the simplest types of data, Likert item data, where people respond with numbers between 1 (strongly disagree) and 5 (strongly agree). A typical distribution is shown in Figure 2.1. For this distribution, the most likely response is 3, with a probability of 0.6 of getting this response.

We cannot really know what the 'true' probability of getting a 3 response is because who exactly is the theoretical population who gives these responses? However, if we have previously administered this item to a group of people, we can make the claim that this is the distribution of responses we have seen previously.

Likert item distributions are relatively simple. However, where data are continuous so that at least in principle they might take any value, such as timing a task, then any specific, precise value has zero probability no matter how large the sample. For such continuous variables, distributions are instead represented by a probability distribution function, such as the classic bell curve shown in Figure 2.2.

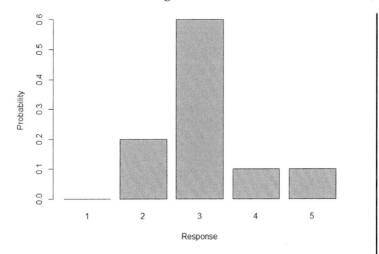

Figure 2.1 A barchart representing the distribution of the responses to a Likert item.

Though this is a very familiar shape, it does not mean the same as the previous chart. The probability of getting a time on task of 20s is not roughly 0.06, being the value of the graph for that time: it is actually zero because getting that *precise* time is incredibly unlikely. To use the graph to find probabilities, take a range of values and find the area under the graph for that range. The area as a proportion of the area under the whole graph represents the probability of getting a value in that range.

Even though such probability distribution functions do not directly represent probabilities, they do help to understand the shape of data. For example, it is clear that the bell curve is symmetric (no skew) and that the most likely times are between 20s and 30s.

Typically, and not always correctly, researchers assume that means are normally distributed (see Chapter 7 for more discussion on this). For the purposes of this discussion, I declare, with a certain omniscience, that the immersion scores can be assumed to be normally distributed under the null hypothesis. This allows us to do a *t*-test (Chapter 11). What this does is compare the difference in means that we measured against the difference in means that would be typical under the null hypothesis that the data is normally distributed. The larger the difference in means, the less likely it is under the null hypothesis. Thus, the *t*-test allows us to move from a specific difference in means to a probability, the all important *p*-value.

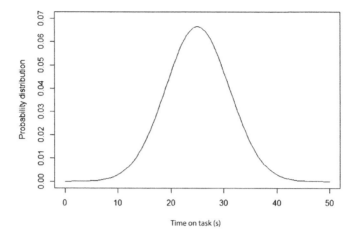

Figure 2.2   A graph representing the normal distribution of the time people take
to do a task.

The *p*-value is the probability that the difference in means we get (or a
larger difference) is likely to have occurred by pure chance under the null
hypothesis. Thus small *p*-values indicate unlikely results *assuming the null
hypothesis*. How small is small? Well, by convention $p < 0.05$ is small, that
is, the probability of getting the data by chance is less than one in twenty.
And that is pretty much the standard level of significance used in HCI.

To summarise, here is the basic argument of NHST:

1. Specify your alternative hypothesis
2. Gather data to test your alternative hypothesis, ideally in an
    experiment
3. Assume the null hypothesis and the distribution of your data under
    the null hypothesis
4. Calculate the probability of getting a difference in means (or
    whatever) by chance alone under the null hypothesis
5. If that probability is less than 0.05 then you have significance
6. Conclude that you have evidence in support of the alternative
    hypothesis.

2.2   Going beyond *p*-Values

Put like this, the NHST seems quite reasonable. And so it is. We have
moved from a general idea to a situation, an experiment, where we have

refined an idea to the point where it can be tested. We then gather data for the test and a significant result is evidence in favour of the idea. This is precisely a type of severe test. And note that in specifying NHST this way, an important semantic move has happened: we have gone from an analysis of the data to support for a (scientific) idea. NHST is very good at addressing the problem of scientific induction (Chapter 1).

I often liken this to a magic trick. The probability of a magician guessing your card by chance is one in fifty-two. This is quite unlikely. The magician then undergoes a sophisticated ritual to make you believe that you are putting her card prediction skills under a severe test: you check the deck is a standard deck; you choose a card without the magician peeking; you signify the card in some special way. The trick works because no matter how elaborate and enjoyable the procedures to ensure a severe test, the magician nonetheless predicts your card. Or better yet, makes it appear inside an unpeeled orange. It is surprising and you know you have been fooled but you do not know how. And in your head, you are definitely not suspecting that on the magician's tour there have been fifty-one other, very disappointed audiences.

In a magic trick, you have moved from the semblance of a fair outcome to a position of very good evidence that you have been fooled. The only difference from an experiment is that all your efforts not to be fooled are genuine. The experiment is a genuinely fair procedure (as far as we can possibly know) and it is only because you had a clear idea of how things work that this seemingly unlikely outcome came about. This should give you some encouraging evidence that your idea is sound.

I am not denying NHST as a scientific, socially constructed practice has problems but when conducted like this, it is every bit as impressive as a good magic trick. And not only does it make good sense, disciplines like psychology and HCI made progress with it. For example, a lot of the provoking and counter-intuitive work on cognitive biases was unearthed using NHST (Kahneman, 2011). This is not to say every researcher using it knew what they were doing, or that no-one engages in *p*-hacking in order to get a paper past reviewers. It is just that as an argument form, it is not inherently mad nor even inherently flawed.

Interestingly, this form of NHST is not advocated by any of its origi-nators: not Fisher, who proposed using probabilities as a way to assess evidence, nor Neyman and Pearson, who went from probabilities to a decision procedure for choosing between explicit null and alternative hypotheses (Gigerenzer, 2004). In fact, the argument form of Neyman–Pearson required an explicit alternative that predicted a clear alternative outcome, for example, that mean immersion would be two points higher under the time-pressure condition, not simply the alternative that the means would be different. What we typically use in modern experiments is a much weaker alternative hypothesis that there is simply some difference in means. However, this should really only be acceptable in the early stages of knowledge when we have no way to predict how big a difference between conditions to expect. In fact, this is actually the kind of situation where p-values are particularly good (Senn, 2001). Over time, our knowledge should grow and so we should produce more concrete alternative hypotheses. Unfortunately in HCI, like quite a lot of psychology as well, we are rarely in a position to predict concretely how much of a difference we should expect to see between experimental conditions. It is just very hard to say because of the complexity, novelty and context of interactive systems.

In modern statistical analysis, then, we use a form of NHST that is an evolution and amalgam of Fisher's perspective and the Neyman–Pearson decision procedure. A fuller discussion of the historical development of NHST and its problems can be found in Dienes (2008, chap. 3). However, if we had stayed using either of the original formulations of NHST, our ability to generate evidence would be weaker (Haig, 2017).

Having said that, NHST, even as outlined above, is a pretty blunt instrument. Essentially, there are two hypotheses in play: the null, which says, for example, that two means are the same and the alternative, which says they are not. Standard NHST produces a p-value on the basis of which we decide whether the alternative is a better explanation than the null. Notice, under standard NHST we are never in a position to say the null is better than the alternative. The method provides no information on that because the statistical calculations are based on the assumption that the null hypothesis is true. Also, there are situations where we want to move from not only testing an idea but also gaining some information about what to expect in future: not just that there is a difference but how much of a difference to expect. This is something that HCI struggles with but this does not mean it should not try (Kay, Nelson, and Hekler, 2016).

What could we do to perform NHST better? As well as using p-values to point us in the direction of evidence, it is useful to generate some understanding of *how much* things are different in the different conditions. That way, in future, we might be able to know what size of difference to look for. The technical term for 'how big the difference is' is called effect size.

A very common mistake is to use the p-value as an indicator of effect size. Only in a very crude sense does a small p-value indicate a big effect.

The *p*-value also folds in the particular test used, and therefore any assumptions you've made about the data, and also the size of your sample. Getting a significant *p*-value is not a good predictor of getting a significant *p*-value in a future, identical experiment (Cumming, 2013).

This in itself does not undermine the NHST argument form. We are not immediately interested in future experiments, we are interested in the experiment that we did conduct as evidence for our ideas. At the same time, if we want our colleagues to replicate our work, something called for in HCI (Wilson et al., 2011), then we should give them some idea of where to look and what to look for.

As also already mentioned, standard NHST does not allow us to give evidence of the null. Another false argument commonly seen (Robertson and Kaptein, 2016b) is that a high *p*-value (close to 1) is evidence of the null being correct. This is simply not the case, because some effects are small and cannot easily be seen in an experiment. That does not mean there are no effects.

To avoid using *p*-values to estimate the size of effects (whether very large or very small), what is needed are more meaningful estimators of effect size. Confidence intervals are one way of estimating effect size and there also are specific statistics that give other useful measures of effect size, for example, Cohen's (Cohen, 1992). These measures are better predictors of effects that might be seen in similar studies and I will give them more careful treatment elsewhere (Chapter 4). What is particularly convenient is that these estimators of effect size are generated quite easily alongside the same procedures that produce *p*-values. In this way, we do not need to overhaul our entire way of thinking about statistics but instead enhance them and move beyond a desperate search for significant *p*-values.

2.3 NHST and Severe Testing

At the end of the day, we want to use NHST as a way to find out new knowledge. Severe testing was developed, in part, to address the problem of induction in science in the face of uncertainty (Mayo, 1996). It is therefore not surprising that NHST, or rather an improved version of NHST, is strongly compatible with severe testing.

The key in both severe testing and NHST is the importance of having an idea to test. In NHST, this is the alternative hypothesis though, in many ways, once we have devised an experiment the experimental hypothesis is actually a statistical hypothesis about how our data behave. What we really want to test is a substantive hypothesis (Spanos and Mayo, 2015) about how the world works. From that substantive hypothesis, we move to a specific aspect of the world in which we can generate a suitable, much more constrained statistical hypothesis and which we can then set about severely testing. This intimately ties together the process of testing the new idea, the substantive hypothesis, with the process of gathering data for the statistical hypothesis. Knowledge, studies and statistics are intimately interwoven.

Earlier in my own statistics journey, I recognised the importance of having an alternative hypothesis as key to guarding against probabilistic anomalies and called this the gold standard statistical argument (Cairns and Cox, 2008, chap. 6). Using severe testing, here is a more refined version of the argument though the essence is the same:

1. Clearly identify the idea about the world that you want to severely test
2. Single out an aspect of the idea that predicts the behaviour of specific measurements, that is, an alternative hypothesis
3. Devise a study that severely tests this idea so that if it is right, it should be obviously right and if it is wrong, it should be obviously wrong
4. Gather data from the test
5. Use statistics to see whether the alternative hypothesis has passed the test, that is, a small p-value
6. Where the hypothesis has passed the test, we have a bit of evidence that the original idea is a good one
7. Where the hypothesis fails the test, we have a bit of evidence that the original idea might not be right

Here I am deliberately vague about what statistics to use except that they lead to producing p-values. Despite the huge debate about p-values, this purely probabilistic argument is still compelling for the same reason that magic tricks still thrill.

At the same time, we are not doing card tricks. We are trying to grow knowledge and knowing the possible vagaries of p-values we should look harder. Confidence intervals and effect sizes give us some idea of how big a difference was seen between experimental conditions. This helps us to understand more about the parameters underlying our hypothesis and ideas and helps us to refine our ideas and predictions in future studies.

Also, where the trick does not work, and p is not low, we can use effect sizes to see whether there is evidence that there is no effect or just a small effect. Narrow confidence intervals around a null hypothesis tell us that any effect is very likely to be small. We can then decide whether it is worth hunting down the small effect, for example, even a small effect in improving a safety-critical user interface could save lives. Or we can decide that for our purposes, it is not important enough and move on with our ideas.

Thus, severe testing in conjunction with considered interpretation of significance testing helps us to be a bit more confident that what we claim to have found is right. NHST has its faults but mostly they arise when we do not think about it carefully enough and engage in ritualistic procedures whether that is as researchers, teachers or reviewers.

2.4 Honesty in Statistics

As you might expect, there are other approaches to generating evidence in the face of uncertain data. The most popular position gaining ground at the moment is Bayesianism, for example Dienes (2008) in psychology and Robertson and Kaptein (2016c) in HCI. I have particular issues with Bayesianism that I discuss in Chapter 3. There are also magnitude-based inference methods that have some resemblance to significance testing but different emphases on interpretation (Schaik and Weston, 2016). And there are other methods, such as likelihood tests, Monte Carlo processes, Markov models and so on, that could be applied but have yet to gain popular interest.

However, all techniques in statistics are limited. We are never able to reason with certainty over the inherently uncertain. I think the problem some of the debates about the 'right' way to do statistics divorce the analysis of data from the generation of data. Useful data are not just lying around waiting to be analysed, despite the claims of Big Data. Where the things we want to understand are subtle and complex, we have to go out and generate the data that help us to understand them better. We need to make experiments that isolate the phenomena of interest and then severely test them. Even then, under severe testing, we isolate only one phenomenon that is one of many potential consequences of the larger, unproven substantive idea about how the world works. A hugely significant, large effect with a substantial likelihood adds only one piece of evidence in support of the idea. And that is regardless of the statistical methods that we use.

NHST has particular problems with an unthinking emphasis on p-values. These have become socialised to the point that there is an unhealthy focus on significance to the detriment of whether what has been found is right or not. However, at the same time, I do not believe any other method would be immune to its own equivalent of p-hacking in the face of such social pressure. We run the risk of throwing out NHST just to replace it with another socially engineered and equally flawed ritual.

Instead, we need to look carefully at the meaning of statistical findings. First, we need honesty in our analysis. As Feynman neatly put it: you must not fool yourself and you are the easiest person to fool (Feynman, 1985, p. 343). And not fooling yourself, with statistics or any method, requires an utter honesty and clarity in your research methods. This is a principle that will keep cropping up in this book because severe tests, unlike magic tricks, require utter honesty.

Interestingly, appealing to that honesty has been proposed as a way to improve psychology research (Simmons, Nelson, and Simonsohn, 2012), not because psychologists are inherently dishonest but because sometimes we forget in our enthusiasm what utter honesty in science is. It is hopeful to believe that a simple appeal for honesty and clarity should bring about improvements in science.

Next, we need to be really clear that no matter how wonderful an experiment is and how delightful its results, it is at best only one piece of evidence. If we stop at one experiment, we are letting our ideas down by not having the conviction to keep severely testing them and so giving them further opportunities to shine. At the end of the day, one experiment just is not enough. The problem is not NHST but science and researchers having sufficient drive to test and refine and test their ideas so that each time they are little bit less wrong.

References

Cairns, Paul and Anna L Cox (2008). *Research Methods for Human-Computer Interaction*. Vol. 12. New York: Cambridge University Press.

Cohen, Jacob (1992). 'A power primer'. *Psychological Bulletin* 112.1, pp. 155–159. (1994). 'The earth is round ($p < 0.05$)'. *American Psychologist* 49.12, pp. 997–1003.

Cumming, Geoff (2013). *Understanding the New Statistics: Effect Sizes, Confidence Intervals, and Meta-Analysis*. Routledge.

Dienes, Zoltan (2008). *Understanding Psychology as a Science: An Introduction to Scientific and Statistical Inference*. Palgrave Macmillan.

Feynman, Richard (1985). *Surely Youre Joking, Mr Feynman*. Vintage.

Gigerenzer, Gerd (2004). 'Mindless statistics'. *The Journal of Socio-Economics* 33.5, pp. 587–606.

Haig, Brian D. (2017). 'Tests of statistical significance made sound'. *Educational and Psychological Measurement* 77.3, pp. 489–506.

Haller, Heiko and Stefan Krauss (2002). 'Misinterpretations of significance: A problem students share with their teachers'. *Methods of Psychological Research* 7.1, pp. 1–20.

Howell, David C. (2016). *Fundamental Statistics for the Behavioral Sciences*. Nelson Education.

Kahneman, Daniel (2011). *Thinking, Fast and Slow*. Macmillan.

Kay, Matthew, Gregory Nelson and Eric Hekler (2016). 'Researcher-centered design of statistics: Why Bayesian statistics better fit the culture and incentives of HCI'. *Proceedings of the 2016 CHI Conference on Human Factors in Computing Systems*. ACM, pp. 4521–4532.

Mayo, Deborah G. (1996). *Error and the Growth of Experimental Knowledge*. University of Chicago Press.

Robertson, Judy and Maurits Kaptein (2016a). 'An introduction to modern statistical methods in HCI'. *Modern Statistical Methods for HCI*. Springer, pp. 1–14.

(2016b). 'Improving statistical practice in HCI'. *Modern Statistical Methods for HCI*. Springer, pp. 331–348.

eds. (2016c). *Modern Statistical Methods for HCI*. Springer.

Rosenthal, Robert (1979). 'The file drawer problem and tolerance for null results'. *Psychological Bulletin* 86.3, pp. 638–641.

Schaik, Paul van and Matthew Weston (2016). 'Magnitude-based inference and its application in user research'. *International Journal of Human-Computer Studies* 88, pp. 38–50.

Senn, S. (2001). 'Two cheers for P-values?'. *Journal of Epidemiology and Biostatistics* 6.2, pp. 193–204.

Simmons, Joseph P., Leif D. Nelson and Uri Simonsohn (2012). 'A 21 word solution'. *Dialogue: The Official Newsletter of the Society for Personality and Social Psychology* 26.2, pp. 4–7.

Simonsohn, Uri, Leif D Nelson and Joseph P Simmons (2014). 'P-curve: a key to the file-drawer'. *Journal of Experimental Psychology: General* 143.2, p. 534.

Spanos, Aris and Deborah G. Mayo (2015). 'Error statistical modeling and inference: where methodology meets ontology'. *Synthese* 192.11, pp. 3533–3555.

Wilson, Max L., Wendy Mackay, Ed Chi, Michael Bernstein, Dan Russell and Harold Thimbleby (2011). 'RepliCHI-CHI should be replicating and validating results more: discuss'. *CHI'11 Extended Abstracts on Human Factors in Computing Systems*. ACM, pp. 463–466.

CHAPTER 3

Constraining Bayes

Questions I am asked:

▷ What are Bayesian statistics?
▷ Aren't Bayesian methods better than significance testing?

Statistics are founded on probabilities and these need interpreting. However, we know that people have very poor intuitions about probabilities (Hardman and Hardman, 2009) and so the sound interpretation of probabilities needs real care. It is all too easy to fall into false conclusions based on p-values (Robertson and Kaptein, 2016a). There are basically two schools of thought on the best way to interpret probabilities and these are called Frequentism and Bayesianism. And they traditionally do not get along very well.

Typically, the Frequentists are those who support null hypothesis significance testing (Chapter 2) and its improvements. They are so called because they (stereotypically) hold that probabilities arise as idealisations of observed frequencies. For example, if you have a fair die and throw it 600 times, the frequency with which each face should come up would be roughly 100 times each. This is idealised to the probability of throwing a one, for example, to be $\frac{1}{6}$.

Frequentists are meant to interpret p-values, therefore, as the probability of getting data, D, from an experiment where the experiment is one of a hypothetical multitude of identical experiments. The p-value is the idealised frequency of obtaining data like this or more extreme assuming the null hypothesis, H_o, to be true. This is sometimes denoted $p(D)$.

Bayesians reject the notion that probabilities represent the frequencies of chance occurrences. Instead, they hold that probabilities represent beliefs in propositions. A Bayesian would hold that the most rational degree of belief that a die will come up with a one is $\frac{1}{6}$ because there is no reason to believe that any one face would appear more than any other. In experimental situations, if we have some alternative hypothesis, H, Bayesians hold that we do not want to know $p(D)$, which would be a degree of belief in the data, but rather the probability (belief) in H given the data, which is usually denoted $p(H|D)$. There should also be some prior belief in H, $p(H)$, and this is updated by data via Bayes' rule to give a more accurate belief:

$$p(H|D) = \frac{p(D|H)}{p(D)} p(H) \tag{3.1}$$

The updated probability (belief), $p(H|D)$, is called the posterior probability because it comes after reasoning about the data. The ratio $\frac{p(D|H)}{p(D)}$ is called the Bayes factor. In a Bayesian world, people may hold different prior beliefs, possibly even due to subjective opinion, but everyone (who is rational) updates their beliefs according to Bayes' rule.

Both views seem reasonable though for some readers the idea that probabilities represent beliefs may seem strange at first glance if only because probabilities are traditionally introduced through the Frequentist approach. Though reasonable, both views do seem to be incompatible, presenting quite different interpretations of what probabilities mean and therefore how they should be used. Furthermore, the sociological problems of NHST that I've briefly outlined in Chapter 2 and further challenges that Bayesians raise that I detail below would both suggest that the Frequentist approach to statistics is fundamentally flawed. If so, we will have to throw out all of our familiar tests and tools of statistics and start again.

That seems drastic. And I would note that plenty of interesting findings have been developed using the Frequentist statistics approach. In fact, in a large study of published research, Wetzels et al. (2011) showed that, across 855 studies using t-tests, p-values and Bayes factors and effect sizes were all broadly in agreement. It would seem that despite some apparently irreconcilable philosophy, both groups tend to agree on what their data mean.

The Bayesian approach is not without its problems either. I am particularly concerned that some Bayesians appear to engage in unsound mathematical reasoning. If you are not doing mathematics carefully, it does not matter what your interpretation is, you are definitely going to come unstuck at some point. Furthermore, specifying prior probabilities ($p(H)$ in Equation 3.1) presents a particular challenge that divides Bayesians and cannot be swept aside.

The problems levelled at both camps though, I would hold, stem from an undue concern for data and not enough for where the data come from. Data rarely magically appear, despite what Big Data people think (Chapter 1), but is sought out and with care. Positioning data properly in the context of the processes that generate the data helps to deflate the challenges that Bayesians level at Frequentists. At the same time, we can acknowledge for certain types of research goals, in particular estimating parameters, a Bayesian approach is very powerful.

A part of me would like to protect HCI researchers from having to worry about such fundamental issues. Why can't we just get on with doing statistical tests? They are hard enough as it is. But if the logical basis for the tests that we want to use is fundamentally flawed, then we need to be looking for other methods to interpret uncertain data. In addition, there are growing moves to use Bayesian methods in HCI (Kay, Nelson, and Hekler, 2016; Robertson and Kaptein, 2016b) and though the rhetoric is strong, I am not sure that it is the full story. Understanding the Bayesian critique of NHST better will help everyone to analyse their data better.

3.1   Defining Probability

At the heart of both the Frequentist and Bayesian views is the notion of probability. What exactly is probability? It is clear that the goal is to exploit mathematics to allow us to do sound reasoning rather than informal qualitative reasoning that happens to have numbers attached to it. This can be seen both in how Frequentists calculate p-values and in Bayes' rule.

Until the start of the twentieth century, much of mathematics was actually founded on quite informal notions, such as what a polyhedron is (Lakatos, 1976) or how to take limits. However, around the start of the twentieth century, as mathematics progressed, it became clear that there were all sorts of odd concepts arising that were challenging traditional mathematical reasoning and this led to various programmes of axiomatisation. Axioms were intended to be self-evident truths in mathematics from which all the rest of the mathematics of a given field could be derived through logic. This appeal for axiomatisation is captured in the progamme of formalisation set out by the mathematician, David Hilbert, in 1920 (Zach, 2006).

The result was that a lot of mathematics in the twentieth century was done to devise suitable axioms for various branches of mathematics and so provide the foundation that Hilbert sought. In probability theory, one particular set of axioms, due to Kolmogorov, was accepted and is still used as the basis for mathematical probability theory today (Grimmett and Welsh,

1989). The axioms are not without controversy because they were somewhat plucked from thin air; in particular, why should probabilities be less than 1? At the same time, they are accepted because they give a useful footing on which to solve problems in mathematics.

The axioms of probability

First, you need to define an event, E, and these collectively give a set of all events $\mathcal{E}$ that you can reason about. An event is something that can have a meaningful probability. Events themselves are based on a set of points, X, because all of modern mathematics goes back to sets. The axioms for events are:

1. $X \in \mathcal{E}$, so that one key event is that something must happen,
2. If $E \in \mathcal{E}$ then $X \setminus E \in \mathcal{E}$, that is, if an event is that E happens, then E not happening ($X \setminus E$) is also an event,
3. If $\{E_i : i = 1, 2, \ldots\}$ is a set of possibly infinite, possibly overlapping events, then their union is also an event. That is, a countable set of events can be treated as a single event.

Having defined events, probability P is a function that maps events to numbers between 0 and 1. The axioms on P are:

1. $P(X) = 1$, usually interpreted to mean that the probability that something happens is 1 and 1 means certainty.
2. $P(E) \geq 0$ for all events $E \in \mathcal{E}$
3. If $\{E_i : i = 1, 2, \ldots\}$ and F is the union of these events, and the events do not overlap, $E_i \cap E_j = \emptyset$, then

$$P(F) = P(E_1) + P(E_2) + \cdots$$

That is, the probability of a set of non-overlapping events is the sum of the probabilities of each event separately.

Intuitively, simple situations where we might like to use probability fit well with these axioms and the calculations based on them give results that accord with evidence. For instance, for the roll of a pair of dice, the set X is a set of pairs of values, one for each die. The set of events $\mathcal{E}$ is made up of any combination of outcomes, for example, the first die is even, or the total of the two dice is 8. Each unique pair of values of each die is assigned the probability of $\frac{1}{36}$ and all other probabilities that we might like to calculate follow. Moreover, the calculations fit the observations of the behaviour of real, fair dice.

One other concept in probability to add to these axioms is the notion of conditional probability, being the probability of an event, A, given that event B has already happened. Note that we are assuming here that A and B are well-defined events, that is, they are in $\mathcal{E}$. It is not hard to show that it is consistent to assign a probability to this event and it is represented as $P(A|B)$, the probability of A given B.

Conditional probabilities have an intuitive meaning. The conditional probability of throwing a six on one die given that you threw a six on another one is still $\frac{1}{6}$ because the events are independent (the outcome of one throw does not and should not affect the outcome of the other). On the other hand, the conditional probability of the next card from the top of the deck being an eight in a game of Gin Rummy depends very much on how many cards have already been discarded and in particular whether other eights have already been discarded. Basically then, $P(A|B)$ intuitively means the probability of A happening now that we definitely know B has happened (though of course in mathematics we do not need to have such intuitive meanings).

3.2   Plausibility

The examples given above all fit the Frequentist model of probability. There is some situation generating a range of possible outcomes (events) and probabilities are the ideal or expected relative frequencies of the events occurring. The main alternative view of probability is the Bayesian view where probabilities express a (subjective) degree of belief in logical statements or hypotheses. For example, Bayesians can meaningfully discuss the probability that the Earth is round as this is their degree of belief in the roundness of the Earth. By contrast, Frequentists cannot easily discuss the probability of the roundness of the Earth because there is no opportunity for repeating the situation where the Earth might turn out to be a different shape. The Earth is round or it isn't! It is hard to see how these two formulations can both be called probability.

Nonetheless, George Polya and R. T. Cox laid the logical foundations for probability and these were comprehensively developed by Jaynes (2003). It is his exposition that is followed here. Accordingly, rather than jump the gun, let's call the degree of belief in logical statements to be plausibilities. The goal is to have a way of reasoning about statements but one where there is uncertainty, so we cannot have a sure way of knowing exactly what the truth is. Plausibility is the degree of belief in a statement and we want to adjust the plausibility of statements as we learn more, say through gathering data in an experiment. Also, we want to be rational, that is, engage in sound reasoning, so for instance where the new data suggest that a statement is more likely to be true, the plausibility of the statement should be increased.

Jaynes recognised that the plausibility of any statement cannot stand alone but must be in the context of some existing, known facts. Thus, all of his plausibilities are in fact conditional on some context, which is often denoted X. What the logical context X actually is is not always fully specified. In some cases, we can be explicit about what X is, in others we cannot. But Jaynes is rigorous in never neglecting this background context, so the plausibility of a statement S is always written as $p(S|X)$. More concretely, if you want to reason about the roundedness of the Earth then you would need to specify the information on which you base your belief. We use small p here for now because plausibilities are not conceptually the same as Frequentist probabilities and I want to keep the two notions distinct. In Jaynes's formulation, it is not possible or desirable to have an unconditional plausibility $p(X)$: that is just not meaningful if you are reasoning rationally.

Additionally, Jaynes is very clear that reasoning is a finite process and that understanding the plausibility of infinite situations, while mathematically interesting, is not crucial to developing a practical theory of plausibility. Where infinity is required, it has to be treated particularly carefully. The result is a set of rules for how plausibility should work. These are given in the next Box.

The rules of plausibility

Unlike the Kolmogorov, axioms of probability, which assign probabilities to elements of a set, plausibility is assigned to logical statements. The logical statements are a finite set of propositions that form what is called a Boolean algebra, which basically means they can be combined using the usual logical operators. For instance, for statements A and B, $A\&B$ (conjunction) is the logical statement that both A and B are the case and $\overline{A}$ (negation) is the logical statement that A is not the case. The further requirement for plausibility is that there is a background context, typically denoted X, possibly made up of other logical statements, in which the plausibility can be calculated.

The rules of plausibility are therefore:

1. $p(A\&B|X) = p(A|B\&X)p(B|X) = p(B|A\&X)p(A|X)$, these two equations together give the Bayes rule of Equation 3.1
2. $p(A|X) + p(\overline{A}|X) = 1$, which means the plausbility of a statement and its opposite add up to 1

These rules look substantially simpler than Kolmogorov's axioms of probability but in fact, because of the finiteness of plausibility and through

careful calculation, it is possible to show that plausibility does in fact match
the laws of probability (Jaynes, 2003, Appendix A). Thus, plausibility, as
far as mathematicians are concerned, is the same as probability. To be
clear though, outside of mathematics the meaning and therefore the use
of probability and plausibility are quite distinct and this is where caution
is needed.

3.3   Unconstrained Bayes

In typical presentations of Bayesian approaches to statistics (Jaynes is not
typical), the emphasis is always on finding the probability of the hypothesis
under consideration in light of the data gathered, not the probability of
the data under the hypothesis. The key equation is therefore what was
presented earlier in Equation 3.1 (Cohen, 1994; Dienes, 2008; Tsikerdekis,
2016), repeated here for convenience:

$$p(H|D) = \frac{p(D|H)}{p(D)} p(H) \qquad (3.2)$$

However, we cannot safely interpret this equation in either a Frequentist or
Bayesian sense.

A Frequentist is comfortable with the probability $P(D)$ because' based on
the assumptions of the variable measured in an experiment or study, there
is a well-defined probability space that models the behaviour of the data.
This probability space is used to calculate the probability of obtaining those
data. The model is usually based on the null hypothesis H_o and assumptions
about the data, but the important factor is that H_o is not part of the events
in the probability space. The null hypothesis is used to shape the model of
events required by the axioms, much as a cake tin is used to shape a cake
but is not itself part of the cake. Because H_o is not in the probability space
itself, $P(H_o)$ is meaningless to Frequentists.

Conversely, many Bayesians are happy to assign a plausibility $p(H)$
to any hypothesis. But note here, the conditional context that is crucial
to Jaynes's mathematical development is missing. This is an important
constraint because it keeps the domain of plausibilities finite. However,
Dienes (2008, p. 89) states that H is free to vary without stating what it
is free to vary over. There is certainly no implication that H is finitely
constrained and that can cause a lot of problems for sound Bayesian
reasoning (Jaynes, 2003, chap. 15).

To make this clear, consider the plausibility of the data $p(D)$ without any context. What is the plausibility of obtaining these data? From one perspective, the data are completely implausible because of all the things that might have been done in the universe, there was no inkling of necessity that these data were plausible. Alternatively, it has to be totally plausible because it is the data that were obtained and that is how the universe has turned out! Without specifying a sensible context, the plausibility of the data is meaningless. More interesting and meaningful is the plausibility of getting these particular data in the context of some understanding about the experiment that generated the data.

In practice, Bayesians often use a Bayes factor rather than do the full calculation of Equation 3.1. The Bayes factor is a comparison of plausibilities of the data under two different hypotheses. For instance, if there are two competing hypotheses, H_1 and H_2, then the Bayes factor having obtained data D is:

$$B_f = \frac{p(D|H_1)}{p(D|H_2)} \tag{3.3}$$

Sometimes, H_2 is the negation of H_1, that is, $\overline{H_1}$, which makes the Bayes factor the relative plausibility of H_1 being correct versus it not being correct.

Bayes factors are used for two reasons. First, in any statistical testing, what is important is which hypothesis is true, typically characterised as whether it is the alternative or the null. Second a key part of Bayesian reasoning is to estimate the prior plausibilities of the hypotheses ($p(H_1)$ and $p(H_2)$), or as we now more properly should say $p(H_1|X)$ and $p(H_2|X)$), which is generally quite problematic (Dienes, 2008; Mulder, 2016, p. 125). How should we set the level of plausibility of two competing hypotheses that does not automatically bias our analysis? For instance, a flat-earther might set the prior probability of the roundness of the Earth to be nearly zero so it would need a massive Bayes factor to adjust the plausibility that the Earth is round to even approach 50:50. Notice also, if the flat-earther refuses any belief in the Earth being round, that is, plausibility is 0, then no amount of evidence will change their mind.[1] Bayes factors avoid the problem of priors by looking at the relative evidence that the data provide for each of the two hypotheses that are being compared. A large Bayes factor indicates the data are making H_1 more plausible that H_2 whereas a Bayes factor close to zero indicates the opposite. No matter how sceptical a prior belief is, the relative evidence stands.

---

[1] This point always makes me think about arguments around global warming…

Here, the logical context, X, is implicit, but not worryingly so, as it is the same for both hypotheses. However, while the move to Bayes factors is sound in terms of comparative evidence (Royall, 1997), it is still final posterior plausibilities that are needed because otherwise it is possible to get very strong statistical evidence for unsound hypotheses. Jaynes (2003) gives a very nice example of how very strong evidence of extra-sensory perception is easily downgraded in the face of a sensible prior (and hence subsequent posterior) plausibilities. Also, for any set of data, however gathered, it is always possible to generate a very special hypothesis H_S that is able to predict precisely the data you obtained (though you would probably need to see the data before you formulated such an accurate hypothesis). No number of alternative hypotheses can give more evidence than this special hypothesis, which is of course nonsense in terms of trying to honestly interpret your data.

There is also a more insidious problem. It can be said that Bayes factors simply talk about the relative evidence that each hypothesis provides. It is therefore possible to consider a multitude of hypotheses, and unlike Frequentist problems of multiple testing (see Chapter 13), the evidence is the evidence regardless of how many alternatives are considered or even if they were generated after seeing the data (Dienes, 2008, p. 133). This is true, but it neglects that for each hypothesis introduced, the plausibilities still must meet the requirement that plausibilities of the hypotheses and their negations add up to one (see Box 2).

If each new hypothesis extends the alternatives to the original hypothesis, then all prior plausibilities need to be updated (Jaynes, 2003, chap. 4). While the evidence for a new hypothesis may be stronger, the posterior plausibility may actually diminish.

One way out of this would be for Bayesians to claim that this mathematical reasoning in the end is just about subjective beliefs and moving them to increasingly sound beliefs so there is no need to carefully meet either the rules of plausibility or the axioms of probability. This is also true but, in which case, why dress up the process with mathematics like Equation 3.1? There are plenty of ways of quantifying belief, such as fuzzy logic (Zadeh, 1988), that do not need to obey such strict rules. And if the reasoning does not fit any accepted version of probability theory, then whatever is being done, it is not mathematics and there are real risks of making unsound conclusions. Not surprisingly, I do not know of any Bayesians who make this claim. At the same time, there are Bayesians who are not sufficiently careful in what they are doing or, at least, in claiming what can be done.

3.4 The Bayesian Critique of Frequentism

Bayesians are of course not irrational and their view of plausibilities leads to some substantial critiques of the Frequentist view. I summarise briefly the main types of problems raised by Bayesians (Wagenmakers, 2007):

1. *p*-values depend on data that were never observed
2. *p*-values depend on possibly unknown subjective intentions
3. *p*-values do not quantify statistical evidence

The first problem is due to an interpretation of the Frequentist position that the *p*-value arises as a relative frequency in a potentially infinite set of identical experiments. Of course, in practice (usually) only one such experiment is done, which means the *p*-value depends on frequencies that are at best hypothetical. An extreme illustration of this is to consider a situation where two experimenters flip a coin to decide which way to run an experiment and hence generate the data to gather. The critique is that, in an accurate Frequentist model, that coin flip should be part of the statistical model, which seems irrelevant to the goal of the experiment. From a Bayesian perspective, this makes no sense because why should a coin flip that otherwise did not affect the data-gathering process influence the analysis?

The second problem is related to this, in that it is about what experimenters might have done but did not. The most common problem is that of optional stopping: sneaking a peek at the data and choosing to stop gathering data when a particular *p*-value is obtained. This is not so idealised. It is seen to be one of the *p*-hacking techniques used to meet the socially required level of significance in an experimental analysis (Simonsohn, Nelson, and Simmons, 2014) (see Chapter 2). The logical problem is that gathering data this way still produces the same dataset as if the experimenter had not peeked and ended up stopping at the same point. The Bayesian view would be that the data still present the same evidence regardless of how it was gathered whereas the Frequentist would have to admit that if done optimally, optional stopping can be guaranteed to result in a significant result (at any level of significance!) undermining the value of the experiment.

The final issue of *p*-values not quantifying evidence has already been discussed in Chapter 2. It is the case that *p*-values are hard to compare between studies and this is why effect sizes are coming in to the Frequentist toolkit (Chapter 4). Of course, Bayesians avoid this problem altogether by using Bayes factors to compare the relative evidence of the data for differing hypotheses.

3.5   Being Careful: A Response to the Critique

As always, when faced with apparently incompatible philosophies held by equally rational and committed researchers, the trick is to keep an eye on what the goal is and also to observe more carefully what these same researchers actually do.

The goal of statistical analysis is to help find evidence in support of ideas in the face of uncertainty. The ideas we want to find out typically in HCI and also other disciplines are not statistical models but things about how the world works. Hypotheses in the Bayesian calculations necessarily have to be related to a statistical model, otherwise there is nothing meaningful linking them to the data, in which case Bayes' rule (Equation 3.1) has nothing to say. Undoubtedly, Bayesian reasoning works extremely well at saying which statistical hypothesis the data best fits. However, as I have noted above, this still needs to be done with caution. It is probably for this reason that many expositions of Bayesian reasoning (Dienes, 2008; Tsikerdekis, 2016) introduce it in the context of parameter estimation. In this sort of study, there is some recognition that an important parameter is needed, for instance the mean reduction in task completion time due to a new and improved user interface. Bayesian analysis allows the researcher to use any and all data, including using priors based on previous studies, to home in on a good estimation of the parameter in question. Kay, Nelson, and Hekler (2016) make precisely this argument in the context of HCI research. Unlike traditional NHST, Bayesian analysis still works even if the actual value of the parameter is the value proposed by the null hypothesis.

However, in many situations in HCI and psychology, as mentioned in Chapter 2, researchers are simply not in a position to estimate parameters or even to find it useful to do so. To take an example from my own work with Alena Denisova, we were interested in seeing what we called a placebo effect in gaming experience (Denisova and Cairns, 2015). Specifically, we were interested to show that telling players about the sophistication of the in-game technology would influence their playing experience, even though the technology did not actually change. It was not really interesting to estimate in advance exactly how much the manipulation influenced player experience. We wanted to see a substantial effect on player experience to give good evidence to our ideas, but we also knew that the effect would vary due to various factors: the game the players played, their knowledge of technology, the quality of our descriptions, the particular aspect of experience measured, and so on. A detailed statistical model is not nearly so useful as having put our idea under a severe test and shown that it survived the test (Chapter 1).

What is happening here is that we are moving from a small, precise statistical hypothesis to a substantive hypothesis about how the world works, of which our statistical hypothesis is just one small facet. And really no amount of probabilistic precision will help with this. It is a matter of argument about how well the experiment was done, how severe a test it constitutes and only then what the statistics say about the data. What the Frequentist probability helps do is not argue about the probability of getting data like this again but instead to argue for the probability that we make a mistake when we claim that our experiment provides evidence of how the world works (Mayo and Spanos, 2011).

One way of looking at this is to think in terms of diagnosing an illness. A doctor may suspect a patient has a liver infection so runs a test on the patient. The test of course is not definitive but it adds to the doctor's belief in the diagnosis. Actually, the precise probability of the test being correct is helpful but ultimately not useful because either the patient does or does not have an infection. The world ultimately is one way or another way and it is not a matter of plausibility or probability. The test informs the doctor but a good doctor would also look for other evidence. As would any good researcher.

Jaynes (2003, p. 139), though clearly a great advocate for Bayesian reasoning, is very clear that the job of the scientist is complicated. He even says:

> ...there is little hope of applying Bayes' theorem to give quantitative results about the relative status of theories.

No matter how well done, Bayesian reasoning (and Frequentist reasoning) alone is not enough to prove (or disprove) theories.

To come back to Wagenmakers' (2007) critique of the previous section, the *p*-values and of course all data depend on values that were never observed. That is the nature of sampling individuals when there is uncertainty. A Frequentist model of data in the severe-testing paradigm is precisely about setting up a test where other data cannot be observed but could have, in principle, occurred. The coin flip example is irrelevant because no aspect of the coin flip adds to the severity of the test constituted by either experiment and therefore to the evidence of the substanative hypothesis. And in any experiment, the second criticism, that the subjective intention of the experimenter is important to Frequentist interpretations, is perfectly accurate but true of all experiments regardless of the school of interpretation. The subjective intention of experimenters is crucial for the sound reasoning that the experiment is in fact testing the idea that it is meant to test. It is beholden on any experimenter to be honest in order to reveal aspects of the experiment that might potentially weaken its severity as a test. The data are not appearing in a vacuum but are the outcome of an attempt to severely test whether an idea is clearly wrong or clearly right. For the third criticism, we agree that some more robust measures of evidence are useful, which is why there is a move in experimental work to report effect sizes (Chapter 4).

The Baysian criticisms of Frequentist statistics, then, are problems about the nature of reasoning over statistical evidence, whereas they are not problems when you move to think of them as being substantive evidence about how the world works. It is just a matter of perspective.

3.6   So, Frequentist or Bayesian?

Based on my own experiences and understanding of how probabilities work and how statistical analysis generates probabilities, my inclination is to be a Frequentist. The NHST approach to testing hypotheses makes a lot of sense to me. However, I try not to be an unthinking researcher who follows the Neyman–Pearson ritual blindly (Chapter 2). And I would not advocate anyone to adopt the ritual Frequentist way. Rather I recommend a more nuanced approach of devising experiments that severely test ideas and use data to reveal how things work rather than simply fooling myself to get published. Some people might prefer to be more Bayesian and I know from reading the work of those who advocate the Bayesian view that they too are very keen not to be fooled by their data.

And that is the key. We need to be honest about what our goals are. Our experiments are probes and tools that help to pry open previously closed things about how the world works. Experiments do not exist in a void but are devised precisely to attempt to reveal the previously hidden. The data they produce therefore speak to that attempt. Our job as good researchers is, with all honesty, to interpret that data to the best of our ability and to say how successful our attempt is.

It is probably for that reason that Wetzels et al. (2011) found broad agreement between p-values, effect sizes and Bayes factors. Good experiments are good experiments and the data speak to that. This is not to say we can be relaxed and careless in our analysis but rather that careful, honest analysis of all types should lead and does lead to similar findings, which is what you would both hope and expect. Remember also that at the end of the day, one experiment is only one small piece of evidence about how the world works. More evidence, and hence another experiment, is better than relying on one statistical result, whether you are a Frequentist or a Bayesian (Wagenmakers, 2007).

References

Cohen, Jacob (1994). 'The earth is round ($p < 0.05$)'. *American Psychologist* 49.12, pp. 997–1003.

Denisova, Alena and Paul Cairns (2015). 'The placebo effect in digital games: phantom perception of adaptive artificial intelligence'. *Proceedings of the 2015 Annual Symposium on Computer-Human Interaction in Play*. ACM, pp. 23–33.

Dienes, Zoltan (2008). *Understanding Psychology as a Science: An Introduction to Scientific and Statistical Inference*. Palgrave Macmillan.

Grimmett, Geoffrey and Dominic Welsh (1989). *Probability: An Introduction*. Oxford University Press.

Hardman, David and David K. Hardman (2009). *Judgment and Decision Making: Psychological Perspectives*. Vol. 11. John Wiley & Sons.

Jaynes, Edwin T. (2003). *Probability Theory: The Logic of Science*. Cambridge University Press.

Kay, Matthew, Gregory Nelson and Eric Hekler (2016). 'Researcher-centered design of statistics: Why Bayesian statistics better fit the culture and incentives of HCI'. *Proceedings of the 2016 CHI Conference on Human Factors in Computing Systems*. ACM, pp. 4521–4532.

Lakatos, Imre (1976). *Proofs and Refutations: The Logic of Mathematical Discovery*. Cambridge University Press.

Mayo, Deborah G. and Aris Spanos (2011). 'Error statistics'. In: *Handbook of the Philosophy of Science*. Ed. by Prasanta S. Bandyopadhyay and Malcolm R. Forster. Vol. 7: Philosophy of Statistics. Elsevier, pp. 1–46.

Mulder, Joris (2016). 'Bayesian testing of constrained hypotheses'. In: *Modern Statistical Methods for HCI*. Springer, pp. 199–227.

Robertson, Judy and Maurits Kaptein (2016a). 'Improving statistical practice in HCI'. In: *Modern Statistical Methods for HCI*. Springer, pp. 331–348.

eds. (2016b). *Modern Statistical Methods for HCI*. Springer.

Royall, Richard (1997). *Statistical Evidence: A Likelihood Paradigm*. Vol. 71. CRC press.

Simonsohn, Uri, Leif D. Nelson and Joseph P. Simmons (2014). 'P-curve: a key to the file-drawer'. *Journal of Experimental Psychology: General* 143.2, p. 534.

Tsikerdekis, Michail (2016). 'Bayesian inference'. *Modern Statistical Methods for HCI*. Springer, pp. 173–197.

Wagenmakers, Eric-Jan (2007). 'A practical solution to the pervasive problems of p values'. *Psychonomic Bulletin & Review* 14.5, pp. 779–804.

Wetzels, Ruud, Dora Matzke, Michael D. Lee, Jeffrey N. Rouder, Geoffrey J. Iverson and Eric-Jan Wagenmakers (2011). 'Statistical evidence in experimental psychology: An empirical comparison using 855 t tests'. *Perspectives on Psychological Science* 6.3, pp. 291–298.

Zach, Richard (2006). 'Hilbert's program then and now'. *Philosophy of Logic* 5, pp. 411–447.

Zadeh, Lotfi Asker (1988). 'Fuzzy logic'. *Computer* 21.4, pp. 83–93.

CHAPTER 4

Effects: What Tests Test

Questions I am asked:

▷ What exactly is an effect size?
▷ Why do I need to use effect sizes? Aren't *p*-values enough?
▷ Which effect size statistic should I use?
▷ How do I show that there is no effect?
▷ Is [*insert test here*] the right test to use for my data?

Null Hypothesis Significance Testing (NHST) puts an emphasis on turning data gathered in an experiment into probabilities, *p*-values. In particular, the lower the *p*-value the better. Within the frame of severe testing (Chapters 1 and 2), the *p*-values are important as they quantify the probability that the data occurred by chance rather than because of the idea that is being severely tested. Low *p*-values give evidence that the idea being severely tested is the cause of the differences seen rather than chance variation. However, the historical and not-so-historical emphasis on *p*-values, to the neglect of any other features of the data, is well recognised to be unhealthy and encourages producing publishable 'findings' rather than advancing knowledge.

Increasingly, applied researchers are making the distinction between statistical significance (*p*-values) and practical significance, that is, how important the research findings are (Grissom and Kim, 2012). When it comes to practical significance, it is no longer simply a case of more is better. For example, my PhD student Frank Soboczenski looked at reducing number-entry errors (Soboczenski, 2014) but error rates in number entry are typically very small: around 2% of numbers are incorrectly entered in any particular experimental context. His experiments therefore struggled at times to capture statistical significance because we were looking for reductions in an already small number. However, if this work is put into the context of entering numbers into a medical device then any reduction in error rates, even if not statistically significant, may be practically important to the design of such devices to prevent deaths due to incorrectly entered numbers (Thimbleby and Cairns, 2010).

Practical significance is usually represented by effect sizes (or more correctly, measures of effect size). In the context of an experiment, an effect size is basically how big an effect the experimental manipulation (independent variable) had on what was being measured (dependent variable). Effect sizes are therefore a form of statistic that is in some ways standardised, like statistical tests, to be meaningful across a range of contexts. Unlike test statistics, and *p*-values in particular, effect sizes are not typically dependent on the number of participants in the study but are intended to represent the general effect on the underlying population of participants as a whole.

In this chapter, I will discuss three approaches to thinking about effect sizes and discuss how different measures naturally associate with different tests. Typical presentations of statistics tend to give the appropriate effect size for a given test. In fact, I emphasise here and in other chapters that if you think about the sorts of effects you would like to investigate first, then the choice of test and measure of effect size would follow from that.

The three main types of effects that I consider are:

1. Location: how much the independent variable moves the 'average' of the dependent variable
2. Dominance: how much the independent variable makes the dependent variable larger in one group than the other
3. Variation: how much variation in the dependent variable is accounted for by the independent variable

After discussing different tests and measures of effect size associated with these effects, I also discuss the importance of realising that in statistics no statistical measure can be taken as the 'real' value, whether that is a mean task completion time or average score on an aesthetics questionnaire. Experiments typically report *the* effect size where, in fact, it is not possible to have a specific size of effect based on samples in an experiment. Instead, around any effect size there needs to be a confidence interval (CI). This has two benefits: first, being realistic about the practical significance of experiments and second, allowing future researchers to synthesise effect sizes across a range of studies.

4.1   Location

Location is a term that basically means where the data are. When a dependent variable is normally distributed for the whole population, the mean of the data is a good measure of location because it is the median and mode and the centre of symmetry in the shape of the distribution. As a concrete example, suppose we were interested in a system used for data-entry in the call centre for new accounts with an electricity company. Skilled call-centre workers would be able to use the system fluently and it might be reasonable to say that, across the range of customer details, customer types and workers, the time it took to set up a new account in this particular system was normally distributed with an average (mean, median or mode!) time to set up an account of five minutes (the solid line in Figure 4.1a). The five-minute figure acts as a useful measure of location.

Suppose now we were to introduce a new system design and we were interested in the effect it had on the speed of setting up accounts. Of course, ideally, we want to reduce the average set-up time. Provided the changes to the system were not too radical, it would be reasonable to assume that the distribution of account set-up times changes wholesale, as shown in Figure 4.1a, to a new mean of 4.5 minutes. More reasonable, though, may be that, at first, the operators are quicker on average but sometimes, due to new features with which they are unfamiliar, then they may also sometimes take longer than previously. This would result in a reduction in location, but also a change in variation (standard deviation) as shown in Figure 4.1b.

Notice also that, in principle, any clearly defined part of a distribution can be used to specify the location, say the upper quartile or the 95th percentile. In Figure 4.1a, all of these values change by the same amount. However, in Figure 4.1b, they change by different amounts, which means their interpretation needs more care. For this reason, measures of location are generally taken to be some sort of average as they (usually!) have a degree of representativeness of the distribution as a whole: five minutes is a reasonably representative task time whereas the 95th percentile value is only typical for the top end of the distribution and not really representative at all of people who are below the mean.

In reality, we cannot easily (or sometimes ever) know the true change in location of the underlying population distribution. So this idealised value is referred to as d_{pop} and in our fictitious example, $d_{pop} = 30$ seconds. In practice, a study is set up to gather data about the effect of the new system and what is calculated is the measure of effect size based on the samples. Such measures of effect size are intended to be representative of the true d_{pop}.

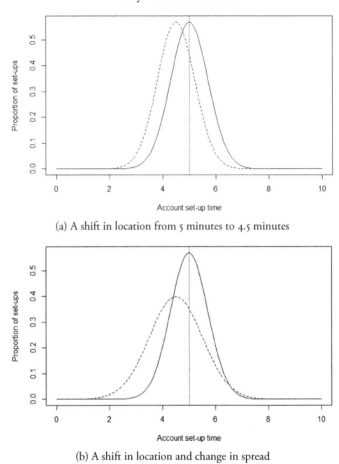

(a) A shift in location from 5 minutes to 4.5 minutes

(b) A shift in location and change in spread

Figure 4.1   An example of change in location in the normally distributed times of an
account entry system.

In the account entry example, the interaction, the time scales and
typical performance are all known. Moreover, the practical significance
of a reduced average task time is well understood; for instance, it might
easily be converted into a Return On Investment (ROI) due to increased
throughput, reduced staff loads and so on. Thus, in this case, measuring
a group of people using the new system and finding that their average
task time was four mins and thirty-eight seconds, gives an effect size of
twenty-two seconds, which could be enough for a manager to make a useful
decision about going ahead with the new system.

In many contexts though, the measure is not so well understood as simply the time on task and, even if it were, there may be little understanding of the typical values of the measure. For example, if we make a change in an archive-search system to reduce users' felt uncertainty, we may not yet have any good idea what typical felt uncertainty is and what would constitute a meaningful change in felt uncertainty. Basically, the d_{pop} is not sufficiently well defined to be meaningful. Instead, the change in location is standardised against the variation in the population, so that instead of thinking about the absolute change in location (mean), the change in location is compared against the natural variation between individuals:

$$\Delta = \frac{d_{pop}}{s_{pop}}$$

Here s_{pop} is the standard deviation of the ideal population. Just as before, the true d_{pop} and s_{pop} cannot be known so both are estimated from the samples in the study. This leads to one of the most well-known and widely used effect sizes, Cohen's d (Cohen, 1992):

$$d = \frac{\overline{X_A} - \overline{X_B}}{s_p}$$

where X is the dependent variable, for example felt uncertainty, and A and B are the two groups of participants, say using the different versions of the system. And s_p is the pooled standard deviation of X based on the standard deviations of both groups of participants.

A d value of 1 corresponds to the difference in average performance being the same as the average difference between two individual participants. This may not mean much to you (yet) but, basically, it would be highly visible in a boxplot or a plot like those in Figure 4.1. Typically, effect sizes are much smaller than this and less likely to be spotted by eye.

If you're familiar with the formula for traditional t-tests, you might notice a striking similarity between the formulae for t and d. In fact, $d = \frac{2t}{\sqrt{N}}$, where N is the number of participants in the study. This gives a useful rule of thumb that when confronted with a t value but no effect size, doubling and dividing by the square root of the sample size gives the effect size.

More generally, the important question is how much of a change in location an intervention causes. To do this, the data from two conditions are compared on the basis of how much some representative value, typically the mean, has moved. This suggests that a t-test and its corresponding d

value is the most appropriate way to test the data. However, t-tests can be unreliable for a range of reasons (Chapter 11) and every problem with data that might make a t-test unreliable is equally problematic for d. Fortunately, so long as something like a t-test can be used and there are corresponding values like d that indicate the effect size of the change of location. For example, if you use a Yuen–Welch test, for reasons suggested in Chapter 11, then d_R (Algina, Keselman, and Penfield, 2005) is the measure of effect size that standardises the change in location for that test.

4.2   Dominance

Though a difference in location has an intuitive and in many cases a practical meaning, in many other situations, particularly in lab-based experiments, we are simply not yet at a stage to know what measures measure and what constitute meaningful changes in location, even when standardised. For instance, if a single Likert item is used to measure the comfort of a digital reading experience (Kamollimsakul, Petrie, and Power, 2014), then a shift in mean comfort score is not terribly useful. What is more useful is whether, on the whole, one digital reading set-up is more comfortable than another. Put mathematically, given two measurements of participants, one randomly chosen from group A and one randomly chosen from group B, the question is what is the probability, P, that X_A is bigger than X_B?

This probability has various names: stochastic dominance, stochastic superiority (Vargha and Delaney, 2000), probabilistic superiority (Grissom and Kim, 2012). I will simply use the term dominance here and in the other chapters. When there is no dominance effect, the probability that a value chosen at random from one group is bigger than a value chosen at random from another is 50:50, that is $P = 0.5$. If every score in one group were bigger than every score in the other (total dominance) then $P = 1$ and, conversely, if every score in one group were less than every score in the other then $P = 0$ (total subjugation?!).

Just as with location, there may be some conceptual ideal dominance of the underlying distributions but practically, dominance has to be estimated through the data gathered in a study. Estimators in statistics are sometimes given a hat, so in this case, the dominance statistic is called $\hat{P}$ (P hat), though in some contexts A' (A prime) is also used (Vargha and Delaney, 2000). This estimator is quite easily, if laboriously, calculated from data by comparing every pairing of scores from the two groups and dividing by the number of such pairs. Fortunately, the Mann–Whitney U statistic happens to be the number of pairs where the score from group A dominates the score from group B, so if the two groups are size n_A and n_B respectively then:

$$\hat{P} = \frac{U}{n_A n_B}$$

You might therefore suspect that the Mann–Whitney test is a test of dominance but sadly this is not quite true. When the assumptions underlying the Mann–Whitney are violated, the test tends to conflate changes in location and dominance. There can be situations where the two changes interfere and the test is unable to detect meaningful differences, no matter how large the sample. This is discussed more fully in Chapter 10. However, tests like the Brunner–Munzel test and Cliff's test overcome some of the problems. They produce estimates of $\hat{P}$ and p-values that indicate the statistical significance of the size of dominance. Thus, where the desire is to look for the dominance of one experimental condition over another, these are robust tests to use because they specifically test for the significance of $\hat{P}$, the size of the dominance.

4.3 Variation

Both dominance and location effect sizes explicitly compare the differences between two groups. When there are more groups, or when considering the effect size of a correlation where there are two variables but no groups, a straight comparison is no longer possible and a different approach to effect size is needed. Instead of thinking about how much one group compares to another, the effect to look for is how much one variable (in particular, an independent variable) accounts for the variation in another variable (typically, the dependent variable). It is easiest to think about this first in the context of correlation.

Suppose we are interested in how the perceived aesthetics of a website improve the time it takes people to complete a task and this is in the context of a utilitarian website, such as a passport-application site. We devise a study that asks a set of participants to complete the task and measure their time on task, T, and the perceived aesthetics, A, of the website. The effect we are looking for is relationship between the two variables and this is done by correlating the two measures (having of course checked for potential problems in the data as discussed in Chapter 14). Suppose the result is $r = 0.4$. This tells us that there is a moderate degree of correlation and in my experience this might just be identifiable by eye in the scatterplot. This, however, is rather vague in terms of effect.

A more concrete interpretation of r is to look at the value r^2. This has the precise meaning that it is the proportion of variation seen in one variable that can be predicted by the other. In the example, $r^2 = 0.16$ so 16% of the variation in the time taken to do the task, T, is accounted for by each participant's perception of the aesthetics, A. Or put another way, $100 - 16 = 84\%$ of the variation in T is not accounted for by A and due to other factors.

Suppose the same effect were studied in an experimental context where there were three designs that differed in aesthetics, being low, medium and high in aesthetic ratings. Now there is no need to measure A as it has already been predefined in the designs, but it is expected that changing A will have an effect on participants' time on task, T, and so increase the overall variation seen in the data. This is called the Proportion of Variance (POV) accounted for by the independent variable. There are several estimates of POV, all with Greek letters: η^2, eta-squared; ω^2, omega-squared; and ϵ^2, epsilon squared. Sometimes they also have hats to show that they are intended as estimators of the 'true' POV. So for instance, suppose the study gave that $\hat{\eta}^2 = 0.06$, then this would mean that 6% of the variation in all the timing data would be due to the version of the site participants had used.

However, all of these measures of variation are strongly associated with underlying models that assume normality and so require the use of the appropriate version of ANOVA (Chapter 12). Because normality is assumed, though these look like measures of overall variation, they are most strongly tied to changes in means and therefore changes in location. Furthermore, it also means that they are undermined by deviations from normality (Chapter 7) and outliers (Chapters 8 and 14). Though there are robust alternatives to ANOVA, these alter the effects that are being examined and hence influence the interpretation of results, as discussed in more detail in Chapter 12.

There is not a simple way around this problem of effects and interpretation of tests when the data cannot be assumed to be normal. The only technique I can suggest is to use a robust test to see if there is any difference between the conditions. This will help to avoid over-testing (Chapter 13). If there is, then compare pairs of conditions using location or dominance effects as you think appropriate to your research question.

4.4   Estimation and Significance

All of these measures of effect size are intended as estimates of the effect on the underlying population of participants, at least theoretically if not practically. The question must then be, how good are they as estimates. The recommended way to indicate the quality of an estimate is to give a confidence interval (CI) for the effect size.

A CI is a range of values of a statistic between which we can be confident (but not certain) that the true value of the statistic for the population actually lies. This definition sits firmly in the Frequentist view of probabilities (Chapter 3): the 95% CI for, say, a Cohen's d estimate, is the range of possible values of d such that if the same experiment were run repeatedly then overall the true value of d, Δ, would lie in that interval 95% of the time. Of course, we still do not know what the true value might be but more pragmatically, because all 95% confidence intervals should most probably include the true value, they should also substantially overlap. Thus, more practically, Cumming (2013) points out that the 95% CI from one experiment should contain the estimate of the same value calculated from a repeat run of that experiment over 80% of the time. In case you are confused why a repeated experiment does not produce a value in the CI 95% of the time, this is because both experiments produce only estimates of the true effect size and neither is likely to actually be the true effect size. Thus, when you combine the error in the first estimate with the error in the second estimate, you arrive at a figure of about 80%.

However, because producing CIs involves the calculation of a probability, the CI is tied to the tests used to generate significance. For instance, a CI for Cohen's d can be calculated using a suitable z-value from a normal distribution (Grissom and Kim, 2012) or, in a more sophisticated way, using a non-central t-distribution (Cumming, 2013). For this CI to be accurate, the assumptions of normality underlying a t-test need to be met and that is not always the case, as discussed above. Robust tests of course can be used and these provide more reliable CIs for their corresponding effect sizes.

Not all tests automatically produce effect sizes, and packages like SPSS certainly do not generate effect sizes or their CIs as a matter of course. By contrast, in R, many of the tests will produce CIs appropriate to the test. For example, the t-test in R produces a CI on the difference in means. Also, Wilcox's implementation of the Brunner–Munzel and Cliff tests of dominance even produce a CI on $\hat{P}$, which is exactly what is required.

The test associated with an effect size does not have to be significant to generate the CI. When the test is not significant, the CI should include 0, that is, there is a reasonable chance that there is literally no effect. Sawilowsky and Yoon (2002) point out, though, that when a test is not significant, the estimate of the effect size is actually misleading because the true value might be zero and small effects can be seen in the data even when there is no real effect in the population. This is the effect-size equivalent of a Type I error (Chapter 9). They therefore conclude, in agreement with others, that effect sizes should not be reported when there is no significance. However, I would argue that this problem arises from only thinking about estimates of effect sizes as *the* value of effect size. If CIs are also thought about and reported alongside non-significant point estimates then the CI will include zero and so not mislead, but also may still include the true value of the effect size, which might not be zero.

Of course, conversely, when the test is significant then the effect size CI should not include zero. This sometimes does not happen when the calculation of effect size does not use the same probability calculation as the corresponding test, for example using a Mann–Whitney test but reporting a $\hat{P}$ CI from a Cliff test. This should be avoided wherever possible as it is not a coherent way to analyse data.

4.5   Big, Small and Zero Effects

Effect sizes are about moving away from statistical significance and thinking about practical significance. Practical significance arises when the effect of one variable on another is big enough to be meaningful. This of course begs the question of what is 'big enough'. There are clearly sizes of effect that in any context are big. For example, in one study I did with MSc student Yuhan Zhou, we looked at the effect of negative or positive phrasing of check-box items. The idea was that checking a box is an affirmative action but if the item is negatively phrased, 'I do not wish to receive emails', then there is a dissonance, at least in English, of affirming a negative. Yuhan set up a very nice experiment to see if checking a negatively phrased item really did cause users problems. It turns out the effect is massive. On average, participants took two seconds to check a positively phrased item and eight seconds to check a negatively phrased one! Note, this is a situation where the change in location alone is sufficient to indicate the effect: a difference of six seconds on a task that only takes two is always going to be of practical importance.

However, more generally, what constitutes big enough is usually a question about when small effect sizes are meaningful. Cohen famously proposed three categories of effect sizes for his *d*: 0.2 is small, 0.5 is medium and 0.8 is large (Cohen, 1992). This basically corresponds to the degree to which an effect is noticeable in a plot of the data, like a boxplot. Small effects are hardly noticeable. But as discussed earlier, even small effects in the context of safety-critical systems could be meaningful.

Reducing errors that could kill people is always going to be of practical significance.

Also, there are contexts where seeing effects is difficult because it is very hard to measure them precisely. David Zendle in his PhD was looking at priming concepts through playing digital games. For example, if people play a game about running mice through mazes, are they then quicker to mentally access concepts related to mice (Zendle, Cairns, and Kudenko, 2018)? Even in ideal contexts of psychology-laboratory studies, priming effects are typically seen as only small changes either in reaction times or scores on standardised tests. David was conducting his work with players playing digital games, not ideal psychological tests, and players were playing online games at home, not in a laboratory. So David's effect sizes were even smaller than expected (around $\eta^2 = 0.005$, which is tiny in most contexts) because of the increased variability in the context of measurement, but they were nonetheless meaningful.

In addition to thinking about the point estimate of effect size, the CI for the effect size also should be considered. It is plausible that the true value of the effect is at the smaller end of the CI. Thus, interpretation of the effect size should not just consider the effect size but this lower limit as well.

At the other end of the CI, it is possible to get a significant result but even so for not only the point estimate but also the upper limit of the CI to be very small. Significance just tells you that that the CI does not include zero as an effect size and, particularly with large samples, the CI can both exclude zero and yet also very narrow so that all the values in the 95% CI are also very small. This can often be seen with large datasets gathered online. For example, in a study, say, looking at gender differences in player kill rates in an online First Person Shooter game, it is feasible to get a large sample of kill rates from 50,000 players. Even a small difference in kill rates between men and women players, say 17.5% versus 18.0% so an effect size of 0.5%, can be significant. However, this is not necessarily an interesting practical significance but of course the true effect size might be larger, so we look at the CI to see in what range of values we might expect the true value to be. However, with such large samples, the CI can be so narrow that even the most extreme effect size in the 99% CI might only be 0.6% and still not of any practical significance.

This leads to consideration of when there might be no effect. NHST is not well set up to find no effect, unlike Bayesian methods. So simply saying that an effect is not significant is not the same as saying there is no effect. What is needed is some way to quantify what an essentially meaningless effect size is. Again, this is a judgement that has to be made in

the context of the domain of study. One researcher's meaningless effect is
another researcher's life-saver. Once a minimum size of meaningful effect
has been decided on, it is possible to reason about how the effect size in
a study relates to that. For instance, the manager of our call centre may
have decided that any improvement in account set-up time of less then ten
seconds is simply not worth investing in.

Two possible approaches are Two One-Sided Tests (TOST) (Grissom
and Kim, 2012) and magnitude-based inference (Schaik and Weston, 2016).
TOST requires the researcher to specify a minimum meaningful effect size.
The TOST then looks to see if the effect size is significantly smaller than
this minimum. If it is then it is reasonable to say that there is no meaningful
(and hence practically a zero) effect. Magnitude-based inference has arisen
in sports science where they are interested in what differences in training
regime bring about a meaningful difference in sporting outcome. The
method similarly specifies minimal levels of useful effect though in a
more elaborate way than TOST and also combines them with a range of
qualitative descriptors that inform interpretation. Both of these approaches
rely on a judgement of what is the smallest effect size that is *practically*
significant, ahead of gathering the data. Because of this, it is possible to
have a result where the conclusion is that there is no practical difference in
a study but there is nonetheless a statistically significant difference. Absence
of meaningful effect is different from absence of any effect.

4.6   Choosing Tests to See Effects

Because of the predominance of p-values in the NHST statistical tradition,
many presentations of statistics first present the tests and then discuss the
measures of effect that go with them. The choice of test is then dictated
primarily by the type of data collected in the study. This leads to discussions
of data type such as categorical, ordinal, interval and ratio. In my early
days of teaching these, I was often bemused by how to use these data types.
Interval or ratio data type alone is not enough to decide to use parametric
statistics and many data types are not purely ordinal (Chapter 18) so the
data types do not really help you to decide between using parametric or
non-parametric tests.

However, if you take effect sizes seriously, then the first question should
be: what effect am I looking for? And based on that, you should then devise
the study that would provide the effect and only then choose the test that
goes with the effect.

For example, the managers of the electricity company are likely to be primarily interested in ROI for the new account-entry system. They need to quantify the average productivity of a call-centre and make decisions based on that. In this context, what is needed is an effect about the change in location of the task times. Further consideration may then need to be made about how robust such effects need to be in the context of what is known about the existing distribution of task times.

By contrast, a café redesigning its website might simply wish to know that more customers find the new design appealing. Website appeal is hard to quantify in any case, so a specific quantity representing amount of increased appeal is not very meaningful. Instead, a measure of dominance of appeal for the new website over the old is more appropriate. This may not help the café owner make a judgement about whether the new website is worth it but, for this sort of redesign, that would always be difficult.

In the context of more basic research in HCI, it can be more complex. Research typically pushes into new contexts, new systems and new interactions. Even for relatively clear concepts like task time, a change of location might not always be the most interesting effect. When the details of typical task times are as yet unknown, measures of dominance may be more appropriate, at least in the first few studies. Other concepts may be well-established but are nonetheless tricky to measure, being subjective experiences quantified imprecisely through questionnaires (Chapter 16). A choice of measure of location, dominance or variation may be a matter of judgement or preference for the researcher. To help bring some clarity to this, I have examined what *can* be found from testing this sort of data. Chapter 18 shows in some circumstances it is really dominance that is the most relevant effect to look for, even if you might have thought location would be suitable.

I am not aware of other people strongly advocating a focus on effects first. In many textbooks, data type is considered most important for deciding what test to use. For this reason, such textbooks might also have a decision tree in the back that says, based on what sort of data you have, what sort of test you should do. My problem with this is that such decisions are based on the idea that you have already collected your data as opposed to having designed a study that generates the data you need.

Stronger arguments about taking data type seriously arise from concerns for the validity if the wrong test is used. Likert items are ordinal and so surely they must be analysed with ordinal methods? *t*-tests and their variations are not suitable even if you are interested in a change of location (Robertson, 2012). The risks of neglecting data type, specifically in Likert items, are highlighted in Kaptein, Nass, and Markopoulos (2010). However, while it is true that Likert items do not generate precise interval data, it is not true that they have no interval-like properties: how else could adding up Likert items lead to meaningful interval data for a questionnaire? Conversely, others would recommend using the best test for the data type to ensure the most power to see effects when they exist, for example MacKenzie (2012, p. 227). This suggests using parametric (location-based) methods over rank-based (dominance-based) methods as they have more power (Chapter 9). But, as discussed in other chapters in this book, you only get more power if the data tested meet the assumptions of the tests used (Chapters 10 and 11). Looking at data type only and not considering the underlying distributions can cause more problems than it solves.

I would therefore make the case that you should be thinking about the effects you expect to see, the design of an experiment to produce those effects and the statistical test that goes with the effect all at the same time. The knowledge we seek, the studies we then devise and the statistics used are intimately connected.

References

Algina, James, H. J. Keselman and Randall D. Penfield (2005). 'An alternative to Cohen's standardized mean difference effect size: a robust parameter and confidence interval in the two independent groups case'. *Psychological Methods* 10.3, p. 317.

Cohen, Jacob (1992). 'A power primer'. *Psychological Bulletin* 112.1, pp. 155–159.

Cumming, Geoff (2013). *Understanding the New Statistics: Effect Sizes, Confidence Intervals, and Meta-Analysis*. Routledge.

Grissom, Robert J. and John J. Kim (2012). *Effect Sizes for Research: Univariate and Multivariate Applications*. 2nd edn. Routledge.

Kamollimsakul, Sorachai, Helen Petrie and Christopher Power (2014). 'Web accessibility for older readers: effects of font type and font size on skim reading webpages in Thai'. In: *International Conference on Computers for Handicapped Persons*. Springer, pp. 332–339.

Kaptein, Maurits Clemens, Clifford Nass and Panos Markopoulos (2010). 'Powerful and consistent analysis of likert-type rating scales'. *Proceedings of the SIGCHI Conference on Human Factors in Computing Systems*. ACM, pp. 2391–2394.

MacKenzie, I. Scott (2012). *Human-Computer Interaction: An Empirical Research Perspective*. Newnes.

Robertson, Judy (2012). 'Likert-type scales, statistical methods, and effect sizes'. *Communications of the ACM* 55.5, pp. 6–7.

Sawilowsky, Shlomo S. and Jina S. Yoon (2002). 'The trouble with trivials (p>. 05)'. *Journal of Modern Applied Statistical Methods* 1.1, pp. 143–144.

Soboczenski, Frank (2014). 'The Effect of Interface Elements on Transcription Tasks to Reduce Number-Entry Errors'. PhD. University of York.

Thimbleby, Harold and Paul Cairns (2010). 'Reducing number entry errors: solving a widespread, serious problem'. *Journal of the Royal Society Interface*, rsif20100112.

Schaik, Paul and Matthew Weston (2016). 'Magnitude-based inference and its application in user research'. *International Journal of Human-Computer Studies* 88, pp. 38–50.

Vargha, András and Harold D. Delaney (2000). 'A critique and improvement of the CL common language effect size statistics of McGraw and Wong'. *Journal of Educational and Behavioral Statistics* 25.2, pp. 101–132.

Zendle, David, Paul Cairns and Daniel Kudenko (2018). 'No priming in video games'. *Computers in Human Behavior* 78, pp. 113–125.

PART II

How to Use Statistics

CHAPTER 5

Planning Your Statistical Analysis

Questions I am asked:

▷ Is [*insert test here*] the right test to use for my data?
▷ I've heard about structural equation modelling [*or other fancy sounding method*]: could I use that to analyse my data?
▷ I've collected my data, how do I analyse it?

Wanting to make sure you are using the right test is a very reasonable concern. Most people are aware that there are wrong tests but that does not necessarily help choose the right test. Unfortunately, very often the person asking me this question has already conducted their experiment and has only just begun to wonder about how to analyse the data. I hope that as you read this essay you are not in the same position, but if you are, I shall tease you with the same story I tell my students, see the Box.

How to get to London?

There was once a city-slicker who hated leaving London but the occasion came one day when he had had to visit a client out in the benighted countryside of Dorset, where roads are narrow, villages are small and signposts are hard to find. He had managed to find his client without any trouble that morning thanks to the joys of GPS but on his way back his phone had run out of charge and he was left to use his wits to get him back to the M3 and the fast route to London.

Not surprisingly, it was not easy. Promising lanes that headed in the right direction steadily meandered off course and one small village looked very much like the next. After about an hour of fruitless wandering down the, admittedly very pretty, lanes of Dorset, our poor traveller was hopelessly lost. He was also very sure that he had seen this particular stretch of country lane at least three times so far.

Just as he was about to give up hope of getting home that evening, he was surprised to see a bus stop that seemed only to serve a particularly large field of cows. Fortunately, at the bus stop stood an elderly lady. Suspecting he had chanced upon a local, he pulled up at the bus stop and winding the window down on his car asked the lady, 'Could you please tell me the way to the M3?' The lady hummed briefly to herself then turned to the man and said 'Well, I wouldn't start from here.'

More seriously, gathering your data without a plan of how to analyse it is fraught with risk. What may seem like a reasonably designed experiment, with well-defined measures to gather suitable data, can suddenly come unglued because of a failure to fit the structures of existing tests. I regularly see students who have more or less attempted a factorial design (only not quite). For example, a typical sort of problem might be when the researcher is interested to see the effect of better menu design on people's ability to find information on a government website. They see that there are two ways to improve menu designs: different wording and different menu lengths. Thus, there are two variables, wording, which that has two versions, and menu length, which is short, medium or long. So far so good: two variables or factors lead naturally to a factorial design. But the researcher is also aware that if the participants were to do all six conditions, they would be fatigued by the end, so they only make five versions of the website recognising that long menus with two different wordings are not so different. The result is five experimental conditions when there are six possible combinations of the two variables. This is nearly, but unfortunately not actually, a factorial design. Of course, there are ways to analyse these data, but they're tricky and nowhere near as clear and clean as a two-way ANOVA. And all of a sudden, their life has got a lot harder. If, moreover, they are PhD students, they may have just given their examiner an excellent line of challenging questioning in the viva.

So the question should not be 'which test should I use?' but 'how should I plan my analysis?' And, of course, that means planning ahead. You would be foolish to plan a holiday in a foreign country without actually knowing how or even if you could get there subject to time and money constraints. Similarly, do not run a study unless you have an analysis planned that could give you the answers you seek. Devise your experiment and your analysis together so that you know when you run the study, there is a sensible analysis that can be done. I have discussed this topic in more detail elsewhere (Cairns, 2016).

In Chapter 4, I propose putting consideration into the effect you expect to see first. Are you in a position to see a change of location or a change of dominance? Or will you need several conditions, in which case you will have to look at a variation effect? But even having narrowed down the options based on the effects, there is still the question of which is the 'right' test to choose. A quick skim through any statistics textbook will reveal a host of tests, some better suited to some effects than others and some that might offer sophistication that you would find useful in your analysis. The problem is that in most textbooks, because they are teaching statistics, the data are often rather magically there for the researcher to analyse. When you are actually doing research and planning a study, precisely what data to gather is still to be determined. Reading a textbook will inform you about tests and what they do, but it will not help you really decide which one to choose: the students who come to me unsure about which test to use will typically have read one or even two such textbooks but still not know what to do.

I cannot tell you which test is best suited to you or your research goals. A decision tree, such as those commonly found in standard textbooks, e.g. Howell (2016), omits the complexity of planning a whole study and the things you do not know about your data before you have collected them. Any alternative decision procedure I offer will be equally flawed. Instead, what I offer here are some principles that will help guide you in evaluating what sort of analysis is best for *you* to do. If you follow these, you may not end up doing fancy statistics but you should be in a good position to do an analysis that you can understand and explain to others.

5.1 Principle 1: Articulation

I would say that the most important aspect of any analysis is that you can explain it, that is, articulate what your analysis means. This is not to say that you can write up the analysis correctly. There are some standard formats for presenting and reporting statistical tests (Harris, 2008). Articulation means being able to take the results of the analysis and explain how those results, first, inform the validity of the alternative hypothesis of the experiment and then inform the idea that was being severely tested in the experiment

There are two reasons that Articulation is so important. The first is that the whole purpose of doing a study is to find out new things and the first person who has to work out what your study has found is you! If you are unsure about what a test is doing then how do you know what it means?

To be concrete, let's take something simple like a *t*-test in an experiment on the times it takes people to set two different microwave cookers to cook for a short cooking time on full power.[1] Like most statistical tests, the *t*-test produces a statistic, the *t*-value, and a *p*-value. When I say, 'it produces' what I really mean is that when a *t*-test is done in a package like SPSS or R, then, typically, the output includes these values and these are the values that you are expected to report at the very least (Harris, 2008). Let's say your favourite statistics package tells you:

$$t(28) = -2.57, p = 0.016$$

What does this mean? Well, it is significant and the *p*-value is nice and small. Great! That should help to get it published but so what?

In this case, being a *t*-test, the test is comparing the means of the times taken by the people using the two different microwaves. The significant *p*-value tells us that those means are significantly different, under the assumption of the null hypothesis. And that those data are normally distributed, though people will tell you that that is not entirely necessary (see Chapter 7).

Thus, an interpretation of the *t*-test needs to be clear that there is a difference in means of times, and more interesting would be to say how big that difference is. The *t*-test alone is not enough and some measure of effect size is needed as well (Chapter 4).

A sound interpretation needs to go further. What might make the obvious interpretation go wrong? All tests are based on assumptions and when those assumptions are violated, the test may be unsound. Or it may not matter. It just depends. The typical threats to a *t*-test come from outliers, deviations from normality, uneven sample sizes and so on (Chapter 11).

All of this needs to be understood and appreciated in the context of your study for you to safely interpret a *t*-test. Yet a *t*-test really is one of the simplest tests available and the terse (or not so terse) output from a statistics package gives solidity to a result that may not be merited. Sound interpretation of statistics is not just about knowing what the statistics packages tell you but also about what might have gone wrong that they did not tell you. Based on this you will be able to articulate to yourself what the test results really mean in the context of your study.

Once you are able to articulate a result to yourself, you need also to think about how to articulate it to your readers. A study is usually intended to contribute to a wider body of knowledge. A test that you

---

[1] See www.youtube.com/watch?v=Bzy5hVvbei8.

cannot explain, even if you understand it, is not much use to the wider community (Abelson, 2012). There needs to be a discussion, usually as a formally labelled Discussion section, where you can explain first what the results mean if they are valid and then discuss why you feel they are valid. Or alternatively you should raise any concerns for statistical validity that you might have. For example, you may have noticed some outliers in the data, or rather people who took a long time to set their microwave, and you want to discuss their impact on the mean times, if any. Once you are clear that the data reflect the behaviour of real people, you might have some interesting ideas about microwave interfaces that you can begin to explain to the reader. Unfortunately, a concern for validity that you cannot resolve may undermine your study but better that the reader knows this and knows why. rather than that the results are treated as rock solid when they are not.

Articulation becomes even harder when the analysis is complicated: *t*-tests are relatively straightforward and many readers will understand them. As long as you can explain what you think is going on, the reader will understand. However, there are many more tests and some are truly complicated, relying both on complex processing of data and accordingly complex interpretation. The challenge these tests present is how to articulate the complexity clearly. This leads to the next principle.

5.2 Principle 2: Simplicity

With modern computers, it is currently very easy to conduct any number of weird and wonderful analyses. A three-way Multivariate Analysis of Variance on four dependent variables with two covariates: no problem! Well okay, maybe a little bit of a problem but really not so hard in a package like SPSS: you just need to know which buttons to press and there are plenty of textbooks and online resources that can tell you that. And usually, a study with a good bit of demographic data, a clear dependent variable and some other possibly interesting measures appears to be suitable for just such a MANCOVA analysis. The problem then becomes, what does it mean? For instance, suppose the MANCOVA comes out with a significant result for an interaction on two of the independent variables but with only one significant covariate. What could this really be telling you?

This goes back to thinking about the purpose of a study. Ideally, a study should be a severe test of an idea (Chapter 1). Some idea of how the world works is distilled down to a specific situation, such as how people set microwaves. When a person uses a microwave, or any other interactive system, there are of course a whole host of things that might be of interest: how long people take to set it; what mistakes they might make; what they feel about the experience; what they think of the product; their preference for different designs; their previous experiences with microwaves, and so on. As good HCI researchers aware of all these potential issues, we might measure these factors in some way, such as with some rating scales. The temptation then becomes to analyse them all as well because after all they might indeed be relevant. In doing so, though, the severe test is lost because now we are interested in any difference, not just some particular, specific difference that links back to the original research idea and that the study is designed to test.

Large, complex analyses, whether many tests of lots of different variables or a large complicated test of everything at once, mean that there are lots of possibly interesting significant outcomes. But which ones constitute the severe test? Is it most important that mean times to set the microwave were different? Or that people who preferred Microwave A tended to make fewer mistakes with it? This is not to say that such large analyses are without value: they can be important to explore issues when moving from a qualitative to a quantitative perspective (Chapter 13). It is just that no idea is being severely exposed to scrutiny.

It is for this reason that I think planning for simplicity is important. If there is a clear idea that is being severely tested then it should be enough just to focus on that one idea and test that. Of course other things might matter but that can be done with auxiliary analyses that support explanations (Chapter 13) or in other studies. At the time of devising a study, make the study and the analysis as simple as possible so that the idea under test has nowhere to hide and this makes the test as severe as possible.

My favourite example of this from my own experience is the work done with Frank Soboczenski (Soboczenski, Cairns, and Cox, 2013). We had an idea that making text harder to read would *reduce* the errors that people make when transcribing the text. This built on some similarly counterintuitive work in psychology about how reading obscured test improved recall of the text (Diemand-Yauman, Oppenheimer, and Vaughan, 2011). The novelty in Frank's work was to see this appear in the context of interactive systems and transcription errors.

Obviously there are lots of ways of obscuring text: colour, font face, emboldening and so on. There are also lots of different styles of text: newspaper articles; research articles; novels; tweets and so on. A first study in this area could have tried to look at all of these. But it didn't need to. If the idea held, any study that showed that some attempt to obscure text resulted in reduced transcription errors would be evidence in support of the idea. So the first experiment that Frank did was to have two colours of text (light-grey and black) and some standard sentences from news websites. Participants saw one colour of text and transcribed sentences for a fixed period. The statistical test was one of the simplest, the Mann–Whitney test, which was suitable for comparing the errors between two groups of participants because we were looking for a dominance effect of people making more errors in one condition than another.

And it worked! Moreover, with this experiment and analysis, there was nowhere to hide. Either reduced quality of text would on average reduce transcription errors or it would not. More complicated analyses would not have been more convincing, in fact, arguably if sufficiently obscure, they could have been a lot less convincing because they would be hard to articulate to the reader.

Of course, such testing is risky. If the effect was very small or did depend on subtle features of the text, then our study could have failed to see this. It is perhaps this exposure to risk that makes simple experiments so convincing when they work. Also, I like to think that, had the experiment not given a supporting result, we would have refined the experiment to see if we could work out whether there ever could be such an effect. At the same time, if the effect was small and hard to see, we might have been wasting our time. As it was, it worked and in the later experiments, we found out a lot more about what does and does not work in such experiments and this became the core of Frank's PhD.

Einstein is supposed to have said that good theories should be 'as simple as possible but no simpler'. That is, if it needs to be complex, it should be but not unnecessarily complex. This is true of statistical analysis as well. Some topics are simply complex: culture and experience cannot be easily controlled in a study, so if that is what is being examined then a more complex analysis is necessary. Also, as ideas develop and become more mature, the questions asked of those ideas also become more sophisticated; for example, how does the readability of the text interact with the complexity of the text's language to reduce errors? In which case, there is no alternative but to conduct more complex analysis. But it is important to bear in mind that the analysis should be no more complicated than the question being asked.

5.3   Principle 3: Honesty

Running a study is a lot of work. It takes time to develop, get ethical clearance, gather the data and analyse them. It is therefore particularly understandable when a study fails to give a significant result that a researcher looks to some way of salvaging something from the work done. Statistics are an easy place to start because they are relatively easy to do once the data is in a statistics package and they might help you find something interesting.

However, there needs to be a principle of honesty. In devising a severe test, there should ideally be only one clear analysis that addresses the idea under test. A clear result is a significant one with an effect that is interesting, or a non-significant result where the effect is almost zero. This way either the idea has passed the test clearly or the idea has completely failed to pass the test when it really should have. Unclear results, where there is some effect but it is not significant, cry out for further analysis to see what might be done. Any such further analysis is no longer severe testing. The idea has already failed and it requires real honesty to admit this, however disappointing that result may be. I regularly read papers where it is clear that if the reported result were the one that was being severely tested then the experiment would have looked quite different. What is being reported is a secondary analysis that has produced the 'all important' significant result necessary for publication (Chapter 2).

This is not to say that further analysis is pointless. Whether or not the result is significant, it is worth exploring the data to see both why something might not have worked but also to be sure that something that seems to have worked has worked for the right reason (Chapter 13). This includes checking the data, such as whether there were outliers (Chapter 8) or whether standard deviations were comparable in size to meet the assumptions of the t-test that was done. These discussions become important for learning from a study, regardless of the significance, and can be beneficial to the reader. But they should be clearly indicated as such to any reader and it serves no one to promote secondary, exploratory analysis to the status of a main finding.

In one piece of work I did with MSc student, Tim Sanders (Sanders and Cairns, 2010), we had a result where adding music to a game significantly reduced players' sense of immersion in the game when in fact we had intended to increase it! This was something of a blow, but careful analysis showed us that most players really did not like the music. It was disengaging rather than more engaging. This was not something we had expected and it had to be reported as a problem for the study. A less honest approach could have been to produce a startling finding that music is less immersive in games. But that is not what we believed. Even though the result was 'wrong', we could learn from it and in the next study in the same paper, we corrected this problem and things started to work more the way we had expected. This is a case where not only did the wrong result prove useful, it actually got published as well. I believe that that is because it added to the understanding of the area. Statistical significance was irrelevant.

5.4 Conclusions

So when it comes to doing an experiment, how should you plan your analysis? I would hold that you should adhere to the three principles:

1. Articulation: choose a test you really understand and can explain to yourself and to others
2. Simplicity: as much as you can plan simpler studies and choose simpler tests because they give better evidence
3. Honesty: stick to the plan however disappointing the results

If you plan studies where the analysis fits these principles, then I think you will be in a good position to help generate new knowledge. The evidence produced by such studies will be easy to understand and unambiguous and so of great value to other researchers. That's all we can ever ask.

References

Abelson, Robert P. (2012). *Statistics as Principled Argument*. Psychology Press.

Cairns, Paul (2016). 'Experimental methods in human-computer interaction'. In: *The Encyclopedia of Human-Computer Interaction*. Ed. by Mads Soegaard and Rikke Friis Dam. 2nd. Interaction Design Foundation. Chap. 34.

Diemand-Yauman, Connor, Daniel M Oppenheimer and Erikka B Vaughan (2011). 'Fortune favors the bold (and the italicised): Effects of disfluency on educational outcomes'. *Cognition* 118.1, pp. 111–115.

Harris, Peter (2008). *Designing and Reporting Experiments in Psychology*. McGraw-Hill Education (UK).

Howell, David C. (2016). *Fundamental Statistics for the Behavioral Sciences*. Nelson Education.

Sanders, Timothy and Paul Cairns (2010). 'Time perception, immersion and music in videogames'. *Proceedings of the 24th BCS HCI Conference*. British Computer Society, pp. 160–167.

Soboczenski, Frank, Paul Cairns and Anna Louise Cox (2013). 'Increasing Accuracy by Decreasing Presentation Quality in Transcription Tasks'. *INTERACT 2013, LNCS 8118*. Springer, pp. 380–394.

CHAPTER 6

A Cautionary Tail: Why You Should Not Do a One-Tailed Test

Questions I am asked:

▷ What does a one-tailed test do?
▷ What is a directional hypothesis?
▷ Should I do a one-tailed test?

It is unusual to go into an experiment without some expectation of how the results should go. And this is appropriate, as it fits with the approach of severe testing (Chapter 1). If an idea is to be severely tested by a study, then the study should have an outcome clearly predicted by the idea. In particular, the idea should predict not only that there will be some changes in the data but also what those changes should be. For example, if a revised website is intended to have a better design, then that should predict an *increase* in customer purchases through the website. Or an improved calculator design should *reduce* the number of calculation errors that people make. Such specific predictions are called directional hypotheses, predictions about whether an experimental manipulation (independent variable) will affect the measured data (dependent variable) by making it go up or down. This contrasts with a non-directional hypothesis, a prediction that the experimental manipulation will have some effect but not say whether the change in the dependent variable will be up or down. Put this way, non-directional hypotheses are somewhat nonsensical: who would make a change to a website design but have no expectation of whether it would make sales better or worse? Of course, a good researcher should be open to the possibility that things have got worse (and good experiments aim not to exclude that possibility) but in terms of providing evidence that the website design is actually better, then worse sales are a bad outcome.

When it comes to testing for directional hypotheses, it is possible to tailor statistical tests to test in a one-tailed fashion, that is, look for directional changes in the data, rather than a two-tailed fashion when there is a non-directional hypothesis. One-tailed tests therefore seem well suited to directional hypotheses and support the goal of severe testing through experiments. In this chapter, I will describe the difference between one-tailed and two-tailed tests but then turn to discuss why it is a good idea, *in practice*, to stick to using two-tailed tests even when you have a directional hypothesis.

6.1 A Tale of Two Tails

As with much terminology in statistics, the term 'tail' was used with reference to the normal distribution and popularised by one of the earliest developers of statistics, Sir Ronald Fisher (Fisher, 1936). However, in terms of how this translates to tests, it is easiest to think about the simple situation of an experiment comparing the daily sales of a new version of a website with an old one and where we can confidently (if implausibly, Chapter 7) assert that this value is normally distributed. As we have assumed ideal conditions, then the appropriate test to compare the mean daily sales is a *t*-test. Regardless of whether the test is one-tailed or two-tailed, the calculation of the *t*-value is the same:

$$t = \frac{\overline{S_{new}} - \overline{S_{old}}}{s_e}$$

where S_{new} and S_{old} are the sets of daily sales figures with the new and old website designs, respectively, and s_e is an estimate of the standard error of the daily sales figures.

A *t*-value is converted to a *p*-value by calculating the probability of getting a *t*-value the same or more extreme than the value obtained from your actual data. The difference between a one-tailed test and a two-tailed test arises from what you mean by 'more extreme'. This is easiest to see graphically, so let's put some concrete values on the example.

Remember that *t*-values also depend on the size of the samples involved so, for the sake of argument, we will assume that the daily sales data were gathered for two weeks with each of the different websites. So with 28 values, the degrees of freedom in this case is $28 - 2 = 26$. Supposing also that for the example, the resulting *t* value is $t(26) = 1.91$.

Figure 6.1 shows the *t*-distribution for 26 degrees of freedom and our specific *t*-value. Usually, people work with a two-tailed test and so values more extreme than this value are ones that are bigger in absolute size, that is, ignoring whether the *t*-value is positive or negative. This set of

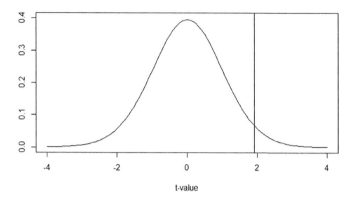

Figure 6.1   The *t*-distribution for 26 degrees of freedom and $t = 1.91$ marked.

values is shown on Figure 6.2a and the area under the curve shown is the corresponding *p*-value. The term two-tailed becomes clear: the more extreme values lie in the two tails of the t-distribution. In this case, the resulting *p*-value is 0.0672.

By contrast, in a one-tailed test, more extreme values are only ones that are literally bigger than the *t*-value. This makes sense in a directional hypothesis because we were never expecting or seeking worse sales from the new website. So more extreme but negative *t*-values that indicate worse sales are a disaster and we don't need to calculate the probability to know that. The set of values for the one-tailed test is shown in Figure 6.2b and again the area under the curve gives the corresponding *p*-value for the one-tailed test. In this case, $p = 0.0336$.

Note in particular, because the *t*-distribution is symmetric, the *p*-value is precisely halved for one-tailed tests compared to two-tailed tests, that is, the same *t* value but half the *p*-value. In this case, a one-tailed test therefore gives a significant result and a two-tailed test is not significant.

6.2   One-Tail Bad, Two-Tails Better

Taking stock of the points made so far:

1.   Directional hypotheses are more focused predictions and therefore more appropriate to the severe testing approach.
2.   One-tailed tests are the appropriate way to test directional hypotheses
3.   One-tailed tests are more likely to see significant effects than two-tailed tests when results move in the direction expected

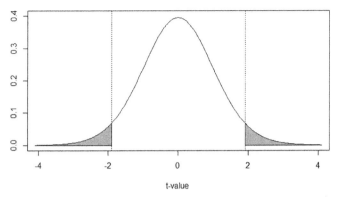

(a) Two-tailed values of *t* more extreme than $t = 1.91$. Shaded area equals the *p*-value of the two-tailed test.

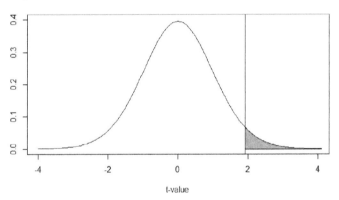

(b) One-tailed values more extreme than $t = 1.91$. Shaded area equals the *p*-value of the one-tailed test.

Figure 6.2   The tails in the *t*-distribution with 26 degrees of freedom.

Put like this, it seems that the right way to do things is to aim for directional hypotheses and one-tailed tests all the time. However, it is the last of the three points that causes problems for one-tailed testing.

For most of the history of the practical application of statistics, it is two-tailed testing that has dominated. It is expected on the whole, in HCI but also in psychology and other disciplines, that two-tailed tests are the norm. This may seem like some arbitrary cultural expectation but remember also that the level of significance, α, is a socially constructed expectation (Chapter 2). Just like driving on the left in the UK, it may not be necessary or even best but it cannot be ignored.

If a researcher has run an experiment but, under the expected two-tailed approach, the result is not quite significant, which of course is typically taken to mean the p-value lies between 0.05 and 0.1, then a one-tailed approach would revise the p-value to be between 0.025 and 0.05 which would make it just significant. And where the focus in a discipline is unduly on whether or not tests achieve significance then this is an important move, particularly in the culture of only publishing significant results (Simonsohn, Nelson, and Simmons, 2014). In the face of the cultural pressures of achieving significance, it might be tempting to say 'Well, my hypothesis was directional all along and so one-tailed testing is the right approach.' Less honourable would be to get a result in the wrong direction from predictions but then revise the hypotheses in the report to give a prediction in the 'right' direction for the results. This would be a particularly dishonest form of p-hacking (Head et al., 2015).

Thus, in the predominant cultural environment of statistical analysis where two-tailed is the norm, then one-tailed testing will look fishy. It suggests to the sceptical reader that the researcher was a little bit too keen to get significance. It might be tempting to argue against this because of our reasonable case for predicting directional hypotheses. However, even in the face of a directional expectation, a good researcher may also feel that perhaps a significant change in the opposite direction should not be over-looked. For instance, work on font sizes for reading digital content suggests, not too surprisingly, that older people find reading larger fonts easier and more preferable (Kamollimsakul, Petrie, and Power, 2014). However, the general consensus is that younger people probably do not have a preference, or may even prefer smaller fonts. In my group, we believe larger fonts are, on the whole, preferred by everyone over smaller fonts particularly on mobile devices. This could have important consequences for designing online content, because it could make everyone a bit happier and have genuine benefit for older readers. Thus, we have done various studies looking at font-size preferences with younger people. However, it may of course be the case that not only do younger people not prefer larger fonts, they may actively prefer smaller fonts. It is because this would be a real and meaningful finding that, even though we have strong directional expectation, two-tailed testing is appropriate and it could also clearly show us if we are wrong.[1] Two-tailed testing does not neglect any result in the wrong direction but actually is testing to see if it could potentially be meaningful in its own right.

---

[1] Though the studies so far, particularly those by my MSc student, Xiaohan Zhou, actually suggest we're right!

In other contexts, it is sometimes, but not often, possible to cleverly set up experiments where two competing theories are in play. If one theory is correct then the results should go in one direction and if the other then the results should go in the other direction. In these situations, a two-tailed test acknowledges the potential dominance of one theory over the other, which is something a one-tailed test could not do.

Thus, two-tailed testing appears to open up the possibility of alternative explanations that, some would argue, are unnecessarily and arbitrarily ignored by one-tailed testing. Of course, there is nothing to stop the one-tailed researcher exploring an unexpected result in the wrong direction. Exploring an unexpected result, though, is not the same as severely testing an idea (Chapter 13) so the researcher also needs to honestly present this rather than trying to angle for the more impressive, but less severe, significant result (Chapter 5).

At the end of the day, the two-tailed paradigm is predominant in statistical testing. The use of one-tailed tests, however legitimate, well-reasoned and well-presented, looks fishy to a sceptical reader. But it is worth bearing in mind that the debate on one- or two-tailed testing shows a fixation on p-values. As Chapter 2 discusses, this is not healthy. A marginal p-value is not a failure if effect sizes are interesting. And relying on a single experiment is bad practice as well. On the whole, it is better to get a marginally significant result with a two-tailed test and do another experiment, than to be accused of, or even tempted into, p-hacking.

References

Fisher, Ronald A. (1936). *Statistical Methods for Research Workers*. 6th edn. Oliver and Boyd.

Head, Megan L., Rob Lanfear, Andrew T. Kahn and Michael D. Jennions (2015). 'The extent and consequences of p-hacking in science'. In: *PLoS Biology* 13.3, e1002106.

Kamollimsakul, Sorachai, Helen Petrie and Christopher Power (2014). 'Web accessibility for older readers: Effects of font type and font size on skim reading webpages in Thai'. In: *International Conference on Computers for Handicapped Persons*. Springer, pp. 332–339.

Simonsohn, Uri, Leif D Nelson and Joseph P Simmons (2014). 'P-curve: A key to the file-drawer'. In: *Journal of Experimental Psychology: General* 143.2, p. 534.

CHAPTER 7

Is This Normal?

Questions I am asked:

▷ Does it actually matter if my data are not normal?
▷ Can I test to see if my data are normal?

It is hard to avoid the historical dominance that normality has in statistics. Despite the fact that statisticians have been persistently refining and improving testing procedures, even in recent times, we still tend to use the same old *t*-tests, ANOVAs and regressions that rely on the assumption of normality. These are our traditional parametric tests. There are seemingly mythical claims that these methods are 'robust' to deviations from normality. In some cases, such as more complex designs involving multiple dependent or independent variables, there does not seem to be an alternative to the old warhorses that rely on normality. But actually, with modern advances in statistics and the computing power to make them practical, there is in fact no need for an unthinking assumption that data are normal (Wilcox, 2010). However, given how focused many researchers are on normality and using parametric tests, here I would like to discuss how we simply do not know when data are normal and that there is no easy way out of this problem because testing for normality does not help with statistical analysis. Of course, this might not matter if our parametric tests still (more or less) gave the right answers but, as I will also discuss, we do not consistently know that either.

7.1   What Makes Data Normal?

Normal data are often equated with the classic bell curve, such as that shown in Figure 7.1, but it is not always clear what that means. The bell curve reflects continuous data, that is data that in principle can take the value of any real number, x. The formula for this curve depends on two parameters, μ and σ, and is given by:

$$f(x) = \frac{1}{\sqrt{2\pi\sigma^2}} e^{\frac{(x-\mu)^2}{2\sigma^2}}$$

The two parameters that give the name to parametric tests are the mean, μ, and the standard deviation, σ. The mean is both the median and the peak of the curve and the curve is symmetric about the mean. The standard deviation is basically a measure of the width of the bell shape. Another feature of the curve is what is often referred to as light-tails: the area under the bell curve at the tails, although in principle representing an infinite range of data values, is even so very small.

The way to understand the bell curve is that the probability of any particular item of data lying in a specific range is given by the area under the bell curve over that range as a proportion of the total area under the curve. Figure 7.1 illustrates that the probability of being within one standard deviation unit of the mean is roughly 67%.

In practice, real data cannot actually fit the bell curve. For a start, real data do not produce a continuous range of real values. Data are always limited by the accuracy of the measuring instrument, for instance tenths of a second on a timer. In HCI, data are often further restricted to whole numbers, such as the number of errors people make or even their response on a Likert item, which is further restricted to be a whole number in a fixed range. In these cases, it is understood that data are not actually normal but can be

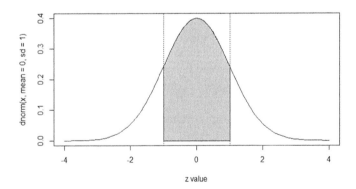

Figure 7.1   The classic bell curve of a normal distribution with mean 0 and standard deviation of 1. The shaded area represents the data within one standard deviation of the mean and corresponds to about 67% of the area under the curve.

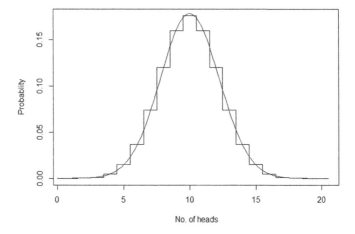

Figure 7.2   A binomial distribution for a fair coin flipped 20 times and
its normal approximation.

reasonably approximated, for the purposes of calculating probabilities, by
the normal distribution.

For example, suppose we flip a fair coin 20 times. The number of heads
gives us a data point. We would expect to get 10 heads in 20 flips, but we
would also not be surprised to get 9 or 11 heads. The probability of getting
a specific number of heads is given precisely by the binomial distribution
but approximately by the normal distribution, as can be seen in Figure 7.2.
The approximation is sufficiently precise that for all practical purposes the
normal distribution can be used to calculate probabilities of getting any
particular number of heads.

Aside from not being actually continuous, many types of data that we
gather are not normally distributed. Error data typically are very skewed
(asymmetrical) with most people making one or two errors and a few people
making several errors and a very few making really a lot of errors. Data
arising from connections between people, such as the number of Friends in
Facebook, citations to papers or links between websites, usually have what is
called a power law distribution (Barabási and Albert, 1999). This is characterised by a very long (heavy) tail and is nothing like a normal distribution.

In my own work also, I have noticed that usability data, such as the time
it takes for people to complete a task, tend to have an outlier (Schiller and
Cairns, 2008). This is something I have reproduced in subsequent student
projects.[1] Such regularly occurring outliers do not fit with the light-tails

---

[1] Adam Auskerin, 2010, unpublished BSc project.

needed for a normal approximation. Outliers are discussed more fully in Chapter 8.

However, normal distributions have come to dominate psychology statistical methods and hence HCI methods, primarily because of a strong mathematical result called the Central Limit Theorem. This result is over 200 years old and originally attributed to Laplace, though with many refinements and improvements made to it subsequently. (See Chapter 1 of Wilcox (2010) for a good historical account.) What the theorem says is that for large enough samples, the sampling distribution of the means is more or less normal. This needs a bit of explanation.

For a given sample of data of a specific size, such as might be collected in a study, the mean of that sample captures a single data point about the whole sample. If that study were repeated, each time gathering a sample of the same size, the means would become data points that could form a distribution. This distribution is called the sampling distribution of the mean and there is a different sampling distribution for each size of sample. What the Central Limit Theorem says is that by choosing a large enough sample, regardless of the original population distribution, the sampling distribution of the mean can be made as close to normally distributed as we want.

It is in fact for this reason that the binomial distribution for flips of a fair coin is approximated by the normal distribution. Adding up the number of heads from 20 flips is logically equivalent to taking the mean number of heads over 20 flips (we just have not divided by the number of flips). A single flip gives very non-normal data: either 0 heads or 1 head. But the Central Limit Theorem says that, regardless of the non-normality of the individual coin flip, for large enough samples, in this case 20, the (mean) number of heads is pretty accurately approximated by a normal distribution.

Better yet, the mean of the sampling distribution of the means converges on the mean of the population. So not only does the sampling distribution become normal, it gives a way to identify the mean of the population, which is usually what really matters when running experiments. So for our coin flips, the mean of the sampling distribution is 10 flips, as you would hope for the mean number of heads in 20 flips of a fair coin.

Thus, it seems that the whole problem of the normality of actual data is neatly side-stepped and the Central Limit Theorem just says as long as samples are big enough then there is not a problem assuming normality of the sampling distribution and that's all you need if you want to use parametric tests to calculate *p*-values.

7.2 The Problems of Non-normal Data

The first obvious problem with relying on the Central Limit Theorem is that it begs the question of how big is a big enough sample size to assume normality of the sampling distribution. The coin flip example demonstrates an extremely non-normal distribution and yet even with a sample size of 20, it behaves well with regards to normality of the sampling distribution. Respected statistics textbooks will often use a sample of size of around 30 as threshold at which assumptions about the data itself being normal are no longer so important. For them, 30 is big enough.

However, this is not the case. Figure 7.3 shows that for samples of size 20 from a log-normal distribution, the normal curve does not approximate the sampling distribution at all well and especially not in the left-hand tail. This is particularly important because the tail is what is used to calculate probabilities in parametric tests. Even with sample sizes of 50, the fit of the means to the normal distribution can be poor (Wilcox, 2010, chap 3). It takes a sample of closer to 100 to be able to get a good fit. The Central Limit Theorem guarantees normality of the sampling distribution of the means eventually but does not say where that will happen with practical, real-world datasets.

A second, more serious problem is that even when the samples are large enough to provide a reasonably normal sampling distribution, the

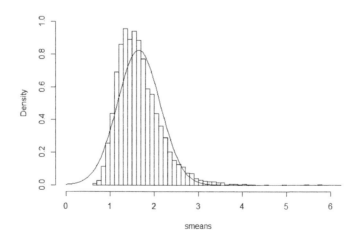

Figure 7.3   A histogram of the means of 10,000 samples of size 20 taken from a lognormal distribution and the normal curve that a parametric test would use to approximate the histogram.

behaviour of the parametric tests may not be what is expected. For instance, the *t*-test relies on an estimation of the standard deviation of the sampling distribution (also called the standard error), which is based on the standard deviation of the sample itself. For a sample from a normal distribution, the errors involved in making this estimation are well understood and accounted for in the *t*-test. But when the sample is not normally distributed, the estimate of the standard deviation can be very inaccurate and so the *t*-test gives faulty probabilities. For example (Wilcox et al., 2013), when the data come from a log-normal distribution, a *t*-test can give a significant result in one direction more frequently than $p < 0.05$ would suggest, that is make a Type I error of seeing an effect when there is none. And in the other direction, because of asymmetries in the sampling distribution, the same distribution may not give a significant result until the true p value is 0.001, that is Type II errors of failing to see a meaningful effect when there is one (see also Chapter 9).

Of course, I have stated that significance testing should not rely on *p*-values alone (Chapter 2). However, the problems of the *t*-test from estimating the standard deviation affect the estimation of effect sizes because they are usually based on the standard deviation of the sample as well (Chapter 4). The problem for effect sizes is worse in some ways because even data that are very close to normal, such as the mixed normal distribution shown in Figure 11.1 of Chapter 11, can have unrepresentative standard deviations. This leads to effect-size measures that indicate very small effects but the effects are in fact substantial in a meaningful way (Wilcox et al., 2013).

A reasonable criticism of these problems might be that they are related to theoretical data distributions and therefore not likely to affect real datasets. This is just not the case. Often there is simply no knowledge of the ways that real data deviate from normality and this undermines the meaning of any statistical testing that assumes normality. Note also, this problem is not a weakness of NHST-related approaches, because Bayesian methods also need to assume a data distribution, e.g. Mulder (2016) and when normal distributions are used the same problems can arise.

7.3   Testing for Normality

If normality of the data itself is so important for allowing the parametric tests to work as expected, then it seems logical to check for normality before doing those tests. There are several tests that can be used to test for goodness-of-fit of a sample of data to a given distribution. These include the χ^2-test, the Kolmogorov–Smirnov test and the Shapiro–Wilk tests. The basic approach would be to use one of these tests to make sure that a sample is normal and then apply the standard parametric tests if it is.

There are two problems with this approach. One problem is about the use of tests for normality and the other is about what they mean in the context of further statistical testing.

Significance tests are well-aligned to the severe testing of hypotheses (Chapter 1). This means that for a statistical test of goodness-of-fit for normality to be useful, the best context would be to have the hypothesis that there is a reason why the data should deviate from normality. This would probably be because there are theoretical reasons to expect that the underlying distribution is not normal. Thus, using such a test to examine whether data are not normal when there is no prior reason for them to be normal or not normal is not a valid severe test. Further, as is well known, absence of significance is not the same as absence of effect (Chapter 4). A null result from a test does not mean that there is not a meaningful deviation from normality. What is needed is a measure of the size of the deviation from normality and, in my experience, this is rarely, if ever, discussed in relation to these tests.

A further problem is that these tests have some rather odd properties. The Kolmogorov–Smirnov test has very low power with small samples (D'Agostino, 1986; Razali and Wah, 2011). This means that even if there were a substantial deviation from normality in the data, for a small sample such as typically seen in an experimental study, say samples of between 30 and 50 participants, the test is not strong enough to detect it reliably. The Shapiro–Wilk test is not much better for detecting non-normal distributions, though for small samples it is better at detecting outliers than the Kolmogorov–Smirnov test (Saculinggan and Balase, 2013). There are alternative tests that perform somewhat better (but still not well for small samples) at detecting both distributions different from normality and the presence of outliers, in particular the Anderson–Darling test (Razali and Wah, 2011; Saculinggan and Balase, 2013) but I am not aware of them being used at all in the HCI literature.

Regardless of whether tests show a significant difference from normality, the second problem is that the tests are not aligned to the problems of applying parametric tests to the data. Failing to detect a deviation from normality does not mean that there is not some deviation from normality that could adversely affect the interpretation of a parametric test. As the previous section noted, even distributions that are very close to normal, like the mixed normal distribution, can have misleading standard deviations and scupper a parametric test. However, the tests for normality would lack the power to detect such distributions. Conversely, even if a deviation from normality is detected by a normality test, it does not mean that the deviation is of the sort that makes a parametric test unreliable.

Testing data for normality in order to apply a parametric test simply does not make a lot of sense because the tests for normality are not sensitive to the features that cause problems to parametric tests. There have been recommendations, which I support, that tests of normality should never be used in deciding if parametric tests are appropriate (Erceg-Hurn and Mirosevich, 2008).

7.4   Implications

All in all, it seems that parametric tests should be abandoned altogether. It is not possible to know when real distributions lead to a good normal approximation of the sampling distribution. It is not possible to test for normality in any way that is, first, reliable and, second, useful for testing. This means it is not possible to know when important violations of test assumptions are occurring in real data. And violation of assumptions can lead to completely misleading analysis both in terms of the *p*-values produced by tests and the effect sizes that go with them.

This sounds like a solid case for the wholesale use of non-parametric tests. However, this would be a premature conclusion. Non-parametric tests are more robust but, despite common belief, non-parametric tests also make certain assumptions about the data that can be violated, just like normality (Chapter 10). Also, there is necessarily a loss of information in the move to non-parametric tests: they often test for the dominance of one group over another rather than look at the change of location between experimental conditions, and it is often actually the change of location (via the means) that is the focus of interest (MacKenzie, 2012). The key is to look more carefully at the purposes of the testing and to choose the right test accordingly. In some cases, it may actually be a parametric test, in others it may be an adaptation of parametric tests (Chapter 11) or it may be non-parametric tests (Chapter 10). It should not be normal practice to assume normality.

References

Barabási, Albert-László and Réka Albert (1999). 'Emergence of scaling in random networks'. *Science* 286.5439, pp. 509–512.

D'Agostino, Ralph B. (1986). 'Tests for the normal distribution'. In: *Goodness-of-Fit Techniques*. Ed. by Ralph B. D'Agostino and M. A. Stephens. Marcel Dekker New York, pp. 367–402.

Erceg-Hurn, David M. and Vikki M. Mirosevich (2008). 'Modern robust statistical methods: An easy way to maximize the accuracy and power of your research'. *American Psychologist* 63.7, pp. 591–601.

MacKenzie, I. Scott (2012). *Human-Computer Interaction: An Empirical Research Perspective*. Newnes.

Mulder, Joris (2016). 'Bayesian testing of constrained hypotheses'. In: *Modern Statistical Methods for HCI*. Springer, pp. 199–227.

Razali, Nornadiah Mohd and Yap Bee Wah (2011). 'Power comparisons of Shapiro-Wilk, Kolmogorov-Smirnov, Lilliefors and Anderson-Darling tests'. *Journal of Statistical Modeling and Analytics* 2.1, pp. 21–33.

Saculinggan, Mayette and Emily Amor Balase (2013). 'Empirical power comparison of goodness of fit tests for normality in the presence of outliers'. *Journal of Physics: Conference Series*. Vol. 435. 1. IOP Publishing, p. 012041.

Schiller, Julie and Paul Cairns (2008). 'There's always one!: modelling outlying user performance'. In: *CHI'08 Extended Abstracts on Human Factors in Computing Systems*. ACM, pp. 3513–3518.

Wilcox, Rand R. (2010). *Fundamentals of Modern Statistical Methods: Substantially Improving Power and Accuracy*. Springer Science & Business Media.

Wilcox, Rand, Mike Carlson, Stan Azen and Florence Clark (2013). 'Avoid lost discoveries, because of violations of standard assumptions, by using modern robust statistical methods'. *Journal of Clinical Epidemiology* 66.3, pp. 319–329.

CHAPTER 8

Sorting Out Outliers

Questions I am asked:

▷ When is an item of data an outlier?

▷ Can I just ignore outliers?

▷ If I see outliers in my data, what should I do about them?

▷ I just removed any outliers: is that all right?

Informally, an outlier, or extreme value, is an item of data that does not fit well with the rest of the data gathered because it is a long way from all of the others. Outliers are known to cause serious problems for statistical analysis (McClelland, 2000). For example, a single outlier can have a strong influence on the mean of a dataset. This has immediate consequences for the use of statistical tests, like t-tests, where the means of samples are the basis for the tests. In fact, it is even a problem for describing data because one outlier in a sample can adjust the mean so much that the mean is not even a sensible representative of the sample or its location. These problems are compounded because standard deviations are calculated based on the mean as well.

Problematic though outliers are, they represent real data gathered during studies. Moreover, like one in a million chances, outliers occur a lot more often in HCI than people expect (Schiller and Cairns, 2008). Any committed quantitative researcher is likely to have to deal with them at some point.

This chapter is not intended to deal with the problems that outliers cause statistical analysis (see instead Chapters 11 and 14) but rather to discuss what exactly causes outliers and therefore how to manage them as part of your data. To achieve this, it is important to have a systematic or at least justifiable way to define and detect outliers.

8.1　Detecting Outliers

The informal definition of an outlier begs two questions. If an outlier is a long way from the rest of the others, where are the others? And what is a long way? A more formal definition of an outlier is that it is a data point that lies outside of the range normally expected of a measure for a particular population (Osborne, 2010). This does not help to address the questions, because we still need to have a specific expectation of what might be a normal range for our data. In a study in HCI done with a new system for the first time in a new context, we are rarely blessed with any clear expectations. To be concrete, a man over 7ft tall is definitely an outlier by height because our common experience tells us that men tend to be between 5ft 6in and 6ft tall and though taller men are somewhat common, much beyond 6ft 6in (the height of most domestic doorways in the UK) is very unusual. There are clear expectations about the heights of men.[1] By contrast, if a study finds that with a new design of App, a person took over 20 minutes to purchase an airline ticket, whilst this seems quite long, we have no expectations nor even intuitions about how long people typically take to do such a task nor whether 20 minutes is unusually long. What is needed is some way to decide on whether the data point is an outlier based on the sample gathered in a study.

As always in statistics, the default approach is with reference to the normal distribution (Chapter 7). The mean represents the average behaviour (or location of the data) and the standard deviation represents the scale of dispersion. This means that transforming the data to z-scores, being the number of standard deviations away from the mean, is a seemingly objective way to see whether a data point is a long way from the others (Field, Miles, and Field, 2012). A threshold is set beyond which a point is declared to be outlying and this is usually motivated by what would be considered rare with respect to a normal distribution. For instance, Tabachnik and Fidell (2001, p. 67) recommend a threshold of $z > 3.29$, corresponding to a p-value less than 0.001.

The use of z-scores to define outliers is widely used in psychology (Bakker and Wicherts, 2014) but it is flawed. Both the mean and standard deviation are calculated using any outlying values and an outlier would tend to move the mean towards itself at the same time as inflating the standard deviation. Both effects tend to reduce the z-score and therefore, effectively, to reduce the apparent severity of an outlier. In fact, Wilcox (2010, p. 32)

---

[1] Though my undergraduates seem to challenge that expectation year on year.

shows how it is possible for an outlier in a small sample to so distort the mean and standard deviation that it completely masks itself, regardless of how outrageously outlying it is. This is extreme but it makes the problem very clear.

What is needed is a similar principle whereby there are robust estimates of location and dispersion that give measures analogous to a z-score but without the circularity of being influenced by outliers. Two methods are commonly seen.

The first, and my favourite, is sufficiently well established to be implemented in a common representation of data, namely the boxplot. This method relies on quartiles: the lower quartile is a value such that a quarter of the data are less than or equal to it and the upper quartile are such that a quarter of the data are greater than or equal to it. The interquartile range, IQR, is the difference between the two quartiles and represents the range of the middle 50% of the data. On a boxplot, the quartiles are the bottom and top of the box, (see Figure 8.1).

The IQR for a normal distribution is roughly $\frac{4}{3}$ of a standard deviation so it works as a useful measure of spread but because it is based on quartiles, it is not altered by one or two outliers. In fact, even if up to a quarter of the data were 'outlying' the IQR would not be affected (though it has to be said that if a quarter of the data are outlying, in what sense are they outliers?). A common threshold then for outliers is a point that is $1.5 \times$ IQR above the upper quartile or below the lower quartile (Emerson and Strenio, 1983). This corresponds to a z-score of about 2.7 in a normal distribution or a p-value below 0.007. This makes outliers identified with this method to be relatively untypical points in a small sample from a normal distribution. Such points are highlighted on a boxplot as separate circles (see Figure 8.1).

A second way to robustly define outliers is using the median absolute deviation (MAD) (Rousseeuw and Croux, 1993) and it is increasingly recommended for its robustness (Leys et al., 2013; Wilcox, 2010). The basic starting point for the MAD is the median of the sample. The distances of each data point from the median (ignoring sign) are called the absolute deviations. The MAD is the median of the absolute deviations of all the data. Sometimes the MAD is multiplied by a fixed factor to make it equal to the standard deviation were the underyling distribution normal. Outliers are then classed as points lying a fixed number of MADs away from the median. Typically 3 is used as this threshold.

The MAD is somewhat like the IQR. Just as 50% of the data lie between the upper and lower quartiles, so 50% are within 1 MAD of the median. The difference is that by definition the MAD is symmetrical about the median

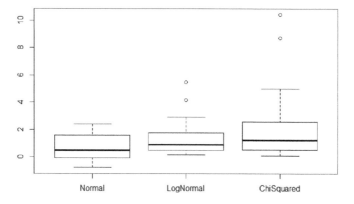

Figure 8.1   The boxplot of random samples of size 29 from each of a normal, lognormal and χ^2 distribution. Outliers are indicated as circles, though because outliers are defined with reference to the normal distribution, these points are not in fact outliers for the distributions they belong to.

whereas the the box on a boxplot can be asymmetric about the median. This has the result that when the underlying distribution is known to be asymmetric about the median, like the number of errors people make in an interaction, the MAD gives a smaller threshold than the IQR and hence more of the data can be classified as outlying. This seems to me to be an artefact of the classification process rather than revealing the structure of the data. For this reason, I prefer the IQR way of classifying outliers as it reflects that distributions can be asymmetric and that outliers may be considered less outlying if they are in the same direction as the skew.

In practice, either the IQR or MAD method provides a robust method for detecting outliers in data measuring a single variable, and the IQR method comes for free with boxplots. Both are preferable to over-relying on z-scores because of the circularity inherent in that outliers influence z-scores. I do not address here ways of detecting multivariate outliers but the same principles apply where robust measures of dispersion are used to say when points are a long way from all the others.

8.2   Sources and Remedies for Outliers

Once outliers have been detected, the question is then what to do about them. This of course depends on why there might be outliers in the first place. The basic causes of outliers in HCI studies can be broken down into the following (Osborne, 2010; Bakker and Wicherts, 2014):

1. Errors in the data
2. Mischievous participants
3. Faulty study design
4. Natural variation

Of course, ahead of time, it is impossible to say which of these is responsible, but if you do have outliers, I recommend working through these alternatives in this order to see if you can discern the cause.

8.2.1 Errors in Data

Whenever data are moved from one format to another, there is always the problem of errors creeping in. Old-fashioned paper questionnaires need to be scored and then the data entered into a useful format for analysis, like a spreadsheet or stats package. During this process, mistakes might creep in. Or there can be simple transcription errors when recording values, such as noting down the score from a game. Whatever the process, there is always the risk of error. Of course, every care should be taken to avoid such errors, but they do occur and sometimes the result is an item of data that appears as an outlier. Impossible or implausible values are a particularly good indicator of this problem, for example, a questionnaire score of 150 when the top score possible is 100 or a time to complete the experimental task of 5 hours when the average time is 5 minutes.

Thus, the first question when faced with an outlier is: is this data value correct? If it is not, then I am afraid that it means a mistake has crept into your data-entry processes and you would be better off checking everything to make sure there are no other mistakes. For example, in identifying why a questionnaire mark is so high, it may come to light that an item has not been reverse scored. In this case, all scores need to be recalculated. The outlier represents an extreme outcome of the mistake but it indicates where other mistakes might lie. And having gone to all the effort of collecting data, it is best to make sure it is of the highest possible quality so, in fact in this case, an outlier is helping to improve your overall data quality.

8.2.2 Mischievous Participants

Anyone who has run an experiment will tell you that participants are funny. Which is another way of saying we're all funny sometimes. Participants may diligently complete all the tasks you ask of them and then have an almost casual disregard for the questionnaire you present them with.

Some participants will check their phone mid-experiment. Others will try to 'game' the system to see what happens. Others will try to guess (and provide) the 'right answers' that you are looking for. It is impossible to predict what people will do, which is partly why we need experiments in the first place.

If the researcher is present during the experiment, participants who are not doing what is expected can be more easily identified but not always. When studies are conducted online, there is no such opportunity and the data must be used to identify participants who have not entered wholly into the spirit of the study. Outliers are a good way to start to identify such participants.

For instance, very high scores on a questionnaire might indicate that a participant went all the way down scoring 5 out of 5 for every item. This may reflect their astonishing enthusiasm for the experience, but where questions have both positive and negative phrasings, it is more likely that they were not trying to answer in any meaningful way.

Another problem might be unfeasibly long task times or a large number of errors. Both suggest that either the participant did not know what they were doing or were not doing what was intended by the researcher.

Outliers reflect extreme outcomes, but if you formulate a hypothesis about how outlying participants were behaving then it would be good to check some other way to see if this is a systematic problem for all participants not just those that appear to be outliers. For example, some people might answer 5 for all questionnaire items, others might answer 3. The 5-scorer appears as an outlier but the 3-scorer probably appears as very average. But the 5-scorer suggests a way to look for potentialy mischievous participants. Even better would be, with an online questionnaire, to see how long participants took to answer the questions. Participants taking 10 seconds to answer 50 questions may not have been reading the questions very carefully.

In all these cases, where outlying scores suggest that participants were not engaged appropriately (for your purposes) in the study, then they and all similar participants need to be identified and removed. If this turns out to be more than a mere handful of people, this might suggest an issue with the study or even a more systematic problem with what the study is about.

8.2.3   Faulty Study Design

Despite the best of efforts, there can be lots of ways in which a well-designed study can end up flawed. I once had an MSc student who was looking at the effect of physiological arousal on people's use of an interactive system. The easiest way to increase arousal is to get people to do some exercise. Unfortunately, in the weeks in which she was running the study, temperatures in London reached an all-time high and exercise was the last thing anyone wanted to do. Her experiment, therefore, did not go well.

What may seem obvious with hindsight is not necessarily obvious until something like an outlier raises a warning. Perhaps a timing procedure was done inconsistently, or a confederate in a study started to interact with participants in a different way. The values of the data are correct but based on an underlying change in the study design and so, effectively, come from a different study.

Another problem can be who participates in your study. Many studies are conducted with an opportunity sample, that is, whoever the researcher can get to take part. Early on, you might recruit a participant who is over 40 but he turns out to be the only over-40-year old in the study and moreover was the fastest participant by far. It may be reasonable to exclude such a participant, not because they were not a good participant but they just do not fit with the rest of the participants. Where such principles of exclusion are devised, then they need to be applied consistently. For example, are there any 35–40 year old participants who should also be considered for exclusion? Or indeed, what is the representative age that you are seeking in your study?

8.2.4 *Natural Variation*

The previous reasons for outliers were essentially that the data did not represent typical behaviour expected of people in the context of the study. However, when such alternatives have been ruled out then the only reason for an outlier might be that that is just how the participant behaved. In my own research, I have noticed that in usability studies, outliers occur pretty much all the time suggesting that there is something about interactive tasks that skew people towards taking longer to complete tasks than might be expected by the 'average' behaviour (Schiller and Cairns, 2008).

As discussed in Chapter 7, many underlying distributions of data have long tails naturally, in particular the number of errors that people make with a task. Using the general standard of a normal, symmetric distribution leads to points being classified as outlying when they are in fact just typical for that sort of data even in the context of your study. This can be seen in Figure 8.1 where samples are taken from log-normal and the χ^2 distributions. The samples are relatively typical but nonetheless appear on boxplots as having outliers. These are of course well-defined theoretical distributions but in practice there may be no good indicators of what underlying distributions to expect in the data from a study. Normality is no more privileged than any other distribution and so using it as the standard for classifying outliers could be misleading.

Concerned that outliers unduly influence statistical analysis, some people systematically omit outliers. However, when the underlying distribution naturally has a long-tail, this removes relevant, representative data from the analysis. When such data are left in, statistical tests, in particular the traditional parametric tests, become unreliable (Chapter 11). Other tests can be used which are not so sensitive to outliers but that can have other consequences both for the quality of the analysis and the interpretation of the results (see Chapters 10 and 11).

I think one good approach is to conduct any analysis first with and then without outliers (McClelland, 2000). If the results give the same story then it is clear that the outliers are part of the same bigger picture and I would report the analysis with the outliers. Where the results differ, things get tricky. However, it should still be considered carefully. If just a few outliers can drastically alter the outcome of a study, then how good is the evidence that the study is providing an answer to your research question? To put this in a more positive light, perhaps outliers that alter the outcomes of a study may be worthy of study in their own right. Exploring the effects of outliers in this way steps back from the main idea being severely tested and instead begins to reveal processes that lead to outyling values (Chapter 13). Nonetheless, the results both with and without outliers may be useful in pushing forward what can be learned about the research question behind the study.

8.3   Conclusions

The key message about outliers is that they cannot be ignored. First, any data need to be checked to see if there are outliers. Second, if outliers are found then they need to be treated in some systematic way. Where the outlier is an accident due to some external influence then all of the data need to be checked for that influence, be it errors, participant behaviour or study design. Where there is no external influence then outliers are valid data points that need to be analysed alongside all the others. At the same time, any analysis needs to recognise the susceptibility of the analysis to outliers and therefore to make sure that they do not carry undue weight compared to other data points. Outliers are not a problem but a fact of research life. They just need to be thought about carefully.

References

Bakker, Marjan and Jelte M. Wicherts (2014). 'Outlier removal, sum scores, and the inflation of the type I error rate in independent samples t tests: The power of alternatives and recommendations'. In: *Psychological Methods* 19.3, pp. 409–427.

Emerson, John D. and Judith Strenio (1983). 'Boxplots and batch comparison'. In: *Understanding Robust and Exploratory Data Analysis*. Ed. by David Hoaglin, Frederick Mosteller and John Tukey. New York: John Wiley & Sons, pp. 58–96.

Field, Andy, Jeremy Miles and Zoe Field (2012). *Discovering Statistics Using R*. Sage.

Leys, Christophe, Christophe Ley, Olivier Klein, Philippe Bernard and Laurent Licata (2013). 'Detecting outliers: Do not use standard deviation around the mean, use absolute deviation around the median'. *Journal of Experimental Social Psychology* 49.4, pp. 764–766.

McClelland, Gary H. (2000). 'Nasty data'. In: *Handbook of Research Methods in Social Psychology*, pp. 393–411.

Osborne, Jason W. (2010). 'Data cleaning basics: Best practices in dealing with extreme scores'. In: *Newborn and Infant Nursing Reviews* 10.1, pp. 37–43.

Rousseeuw, Peter J. and Christophe Croux (1993). 'Alternatives to the median absolute deviation'. *Journal of the American Statistical Association* 88.424, pp. 1273–1283.

Schiller, Julie and Paul Cairns (2008). 'There's always one!: Modelling outlying user performance'. *CHI'08 Extended Abstracts on Human Factors in Computing Systems*. ACM, pp. 3513–3518.

Tabachnik, Barbara G. and Linda S. Fidell (2001). *Using Multivariate Statistics*. 4th ed. Allyn and Bacon.

Wilcox, Rand R. (2010). *Fundamentals of Modern Statistical Methods: Substantially Improving Power and Accuracy*. Springer Science & Business Media.

CHAPTER 9

Power and Two Types of Error

Questions I am asked:

▷ How many participants do I need?
▷ How can I calculate power for a study that has never been done before?
▷ What does it mean to say one test is more powerful than another?
▷ What does it mean to say a test is robust?

The purpose of an inferential statistical test, broadly speaking, is to help us work out whether a difference we see between groups of people (or data) is a real difference or just chance variation. Real differences are ones that would belong to the population in general and so appear automatically in any samples of that population, whereas chance variations are just differences particular to the samples of data we have collected. The power of a test is a measure of how reliably the test can detect differences when they are real.

To illustrate this concretely, suppose there are two different interfaces that have been developed for a system that allows people to register to vote in general and local elections. If one interface really is quicker to use than the other across the population of people who need to register to vote, then the power of a test is an indication of its ability to detect which is quicker based on samples of people who use the different interfaces.

However, if you think carefully about this in the context of research studies, we do not typically know what is real before we go into the study. How do we know which interface really is quicker for voters to register until we try it out? A severe test is based on the *belief* that something is real but the study is designed to reveal whether it is really real (Chapter 1). How then can we ensure the power of a test when, for all we know, there may be no real difference to be seen, no matter how powerful the test is?

This makes power something of a tricky concept and might explain why it is not often explicitly considered in HCI studies (Robertson and Kaptein,

Table 9.1. *The relationship of a test decision to the real world.*

	Test result	
Reality	**Significant**	**Not significant**
No difference	Type I error	Correct
Real difference	Correct	Type II error

2016). The purpose of this chapter is to explain more carefully what power is and to describe two particular ways in which power is used in statistical methods. The first way is quite commonly seen: power is used to calculate sample sizes and so helps to improve the quality of a study. The second way is more about comparing tests and understanding when they work as they should and when they do not. In this second context, power is a measure of the quality of the test itself rather than the quality of the study. Both of these uses of power are related to the probability of misinterpreting the results of a statistical test, that is, to the two types of errors that can be made in statistical testing.

9.1   Type I and Type II Errors

As there can never be certainty about typical HCI data, there is always a chance that the result of a statistical test is wrong. Focusing only on significance gives the commonly seen Table 9.1, where there are two distinct errors possible.

The naming of the errors as Type I and Type II is unhelpfully uninformative. It was only after about ten years of teaching statistics that I started to get them straight in my own head. The terms arose very early on in statistics because of the emphasis of Neyman and Pearson on statistical inference as a decision process (Neyman and Pearson, 1933). In that context, they were originally called 'errors of the first kind' and 'errors of the second kind', which is not any better.

Type I errors are where the test declares a significant result when there is no actual effect in the underlying populations, that is when the null hypothesis holds. In the voter registration system, this would mean that on the whole both interfaces are equally quick to use but a test with a particular sample gave a significant difference between them. Thus, a Type I error is a problem for finding out what is real, because such an error means we are led to support the idea that was being severely tested when in fact it does not hold. In principle, this problem is managed by setting the level of significance, α, at a suitably low level, typically 0.05. This means that Type I errors of finding something significant when there is no real difference are only made 5% of the time. However, Type I errors are a serious sociological problem for science. Scraping in at significance has become so crucial for publication success that it has led to a distortion of scientific practice, called p-hacking (Chapter 2), and the proliferation of Type I errors. In ideal circumstances, the probability of making a Type I error is α but in practice, across published research, it is likely to be much higher because of these problems (Simmons, Nelson, and Simonsohn, 2011).

Type II errors are where a test declares non-significant result when in fact there is something going on, that is, when the null hypothesis does not hold. For the voter-registration system, this would correspond to there actually being a difference in speed of registrations over the population as whole using the two interfaces. A Type II error could occur because in the particular sample no significant effect was seen. This can be just as much a problem for finding out what is real as we may pass over a correct idea in the mistaken belief that it is not real because our experiment did not find it. The probability of making a Type II error is β; however, unlike α it cannot be set by the researcher but arises from the context of research, as will be seen.

Both of these types of error perhaps reinforce the principle that no matter what the result, doing another study is a good idea. A result may be significant because you were just 'lucky' and in fact it is a Type I error or a non-significant result because you were 'unlucky' and made a Type II error. A second study should add more evidence either way.

9.2   Defining Power

The power of a study is its reliability in seeing a real effect. Specifically, it is the probability that the statistical test gives a significant result when there is a real effect influencing the data. Thus, power is the probability of *not* making a Type II error and so power is equal to $1 - \beta$.

Because β is built up in the NHST tradition, it and power can be interpreted in a Frequentist way (Chapter 3). In particular, if the power of a study is 80% then this means it will detect the real effect that it is set up to detect 80% of the time. For instance, if we run 100 identical studies to see if our voter registration systems really are different in registration speed, then in about 80 of those studies we should see a significant effect.

Furthermore, if the power of the study is exactly 80%, then in the remaining 20 of the studies we should fail to see the effect and get non-significant results. Of course, nobody runs identical studies 100 times (in fact, hardly anyone ever even runs identical studies twice (Hornbæk et al., 2014)) so this Frequentist interpretation, like Frequentist interpretations of *p*-values, is purely hypothetical.

Although power is hypothetical, it can be calculated based on decisions about the design of the study:

1. the value of α
2. the test being used to analyse the data
3. the expected effect size
4. the number of participants

Technically speaking α is a decision of the researcher, though one that is hardly ever explicitly made. Nonetheless, power is clearly related to α because how powerful a test is hinges on when a result is declared 'significant' or not. If the level of α is lower so that results need to be more extreme before being declared significant, then power goes down as well because it is harder for any result, even real ones, to be declared significant. Conversely, if we are more relaxed about α then a test result is more likely to be correctly declared significant when there is a real effect (and where there is not) so power goes up.

The test used to analyse the data of a study is usually given by the design of the study, the effect being investigated and the assumptions on the data produced by the study (Chapter 4). A two group between-participant design where the data are assumed to be normally distributed means a *t*-test should be used. Different tests have different power in the same circumstances and in general it is best to use the most powerful test available. It is through the choice of test that the assumptions on the distribution of the data are factored into the power of a study and hence allows the theoretical calculation of power instead of relying on large-scale replications of identical studies.

Typically, the effect size is not known for any particular experiment because if it were it is unlikely that the experiment would be necessary. The effect size is therefore estimated either through comparison with similar experiments or in terms of what is interesting to the researcher. Bigger effects are easier to see, which means power is higher when effects are expected to be large.

Finally, the number of particpants, or sample size, is the only factor that is entirely in the control of the researcher. As a general rule, bigger samples mean more power, but it is more complicated than that.

9.3 Power and Sample Sizes

Obviously, if you are trying to find out what is real and what is not through experiments, then more power is better. Of the factors that can influence power, it is clear that sample size is essentially the main tool that researchers can use to increase the power of a study. Thus, if you desire a particular level of power then, by reversing the calculation that defines the power of a study, it is possible to work out the size of sample that is needed to achieve that power, when the other factors are already fixed. For example, if you are planning a between-participant experiment with two conditions that will be analysed by a t-test, then you might think it is plausible to expect a medium effect corresponding to Cohen's $d = 0.5$ (Chapter 4). The general recommendation for good power is around 0.8, that is $\beta = 0.2$ (Cohen, 1992), to give a good chance of seeing an effect if there is one. In which case, a power calculation (for instance, using the `pwr` package in R) suggests that we need 63 people in each group, that is, 126 participants! Even medium-size effects need large samples for good power. Notice, however, that even with 126 participants, there is a plausible chance (1 in 5) of failing to see the anticipated real effect.

Alarming though sample sizes might turn out to be, it does support the practice of doing a power analysis *before* running any experiment. It would be better to know in advance that a large sample is needed rather than finding out after gathering data that there was a less then 50:50 chance of seeing even a reasonable size of effect and your null result is just as likely to be a Type II error as evidence against your alternative hypothesis.

'How many participants do I need?' is a common question of new researchers. In my experience, it does not help to answer 'How long is a piece of string?', no matter how reasonable a response I think that is. Power calculations, or more accurately sample size calculations, appear to give a principled (and much more useful) way of answering the question. However, at the heart of the calculation is a hole: what is the effect size that is expected? There is simply no answer to this, because if we really knew that we would not need to do the research. What an experiment does is essentially admit the possibility that the real effect size might be zero and that has to be a possible outcome for a fair study.

However, in some circumstances, it may be possible to make an educated guess. David Zendle ran priming experiments in digital games (Zendle, Cairns, and Kudenko, 2018) using measures of priming that had an established pedigree in psychology research. From the psychology studies, we knew that even in ideal conditions priming effects were small, and so in the context of digital games played online we made a pragmatic decision to expect at best half the typical effect size. This made the sample sizes required enormous but then because the games were played online, gathering large numbers of players was not quite as hard as it might have been. Similarly, Joe Cutting in his PhD research looking at pupil-dilation in relation to estimating immersion in games knew that if pupil dilation effects were too small then they would not be useful so there was no point in looking for small effects. In both cases, there were reasonable ways to estimate what effect sizes we would want to look for and hence it was possible to make a plausible calculation of suitable sample sizes.

It is important to reiterate, though, that such calculations are made with reference to specific tests and that these tests fold in their assumptions into the calculations. If a sample size is calculated with reference specifically to the t-test but the data do not meet the assumptions of that test then the nominal level of β could be drastically different from the actual Type II error rate. For instance, data where outliers occur more frequently than expected in a normal distribution can drastically reduce the power of a t-test (Zimmerman, 1994). Thus, the Type II error rate is much higher than the calculated level of power suggests, which means samples need to be bigger still to be sure of seeing effects.

The alternative is to consider other tests and so other measures of effect sizes but, as discussed elsewhere (Chapters 10 and 11), this changes the meaning of the analysis and there can be other assumptions that also need to be considered. In particular, some tests, like the Mann–Whitney test, are only able to detect certain sorts of differences. Where the underlying populations do not meet the assumptions of the test, it is possible that no matter how large the sample is, no study could guarantee reliably detecting the differences. See Chapter 10 for more details of this problem in non-parametric tests.

It seems then that sample-size calculations are built on assumptions and guesses. Moreover, being focused on the two types of error, it would suggest that the only thing that matters is significance, which of course is not really appropriate in modern analysis (Chapter 2). Why then bother with sample-size calculations at all?

I think it is worth stepping back from the problems to think about what a sample-size calculation involves. First, it requires thinking about the design of the study ahead of running the study, and, in particular, thinking about the choice of statistical test before gathering any data, which I think is a good idea (Cairns, 2016). More importantly, though, it forces explicit consideration of both the nature of the effects you are looking for (location vs dominance) and their subsequent sizes. For your particular study, what effect sizes might you expect? What would be the best case? What would be the smallest useful difference you would like to see? What could your study detect? By thinking about these issues, even if the calculation is heavily embedded in black-and-white significant or non-significant results, it fits well with modern approaches to statistical analysis by focusing up front on effects (see Chapter 4 for further discussion).

Of course, the calculation could be based on false assumptions such as normality. However, in almost all cases, the assumption of normality is the best-case scenario. Thus, a sample-size calculation based on, say, a t-test or ANOVA, will give you the smallest useful sample size for a particular level of power. And if you know the assumptions of a t-test will not hold, then you will typically need to move your sample sizes upwards. How much bigger the samples would need to be is hard to say but if your sample-size calculation based on a t-test says you need at least 70 people for power of 80% then running a study with 40, regardless of your assumptions on the data, is likely to be underpowered.

One thing that is strongly not recommended is calculating the power of a study *after* the analysis has been done. This seems to be answering the question 'how powerful was my study, given the effects it identified?' However, this is not a useful thing to calculate (Hoenig and Heisey, 2001; Yatani, 2016). Any study that produces a significant result clearly had enough power to detect the result. And power in itself is not an indication of effect size because even a high power design can lead to a Type II error, that is, no effect being found. The best thing to do once a study is run is to use the estimated effect size seen in the study to calculate a suitable sample size in future studies.

9.4　Power and the Quality of Tests

In modern statistical methods, a lot of attention is paid to the quality of tests: how suitable are tests to use in different situations? Or more pragmatically, how robust are tests to violation of underlying assumptions? A key method for understanding the quality of tests is to understand their error behaviour. A good quality test does not make Type I errors more than expected and yet maintains power (avoids Type II errors) to see effects that are real. In fact, it is the behaviour of tests in relation to Type I and Type II errors that lead to claims like 't-tests are robust to deviations from normality'.

To assess the quality of a test in this way, we need to abstract away from the specific features of a study that lead to particular levels of power and to reason hypothetically about how a test behaves when faced with different sorts of data. Of course, tests are developed mathematically and so their error behaviour in ideal theoretical situations is built-in on condition the assumptions in the test are met. What is important is understanding how tests work when those assumptions are not met, and this can only be done empirically.

Because power depends on α, the first thing to check is that α matches the actual rate of Type I errors. This may seem like a contradiction in terms. Consider then the situation of using a between-participants t-test when data are not normally distributed, say they come from a heavy-tailed distribution like the log-normal distribution (Wilcox, 2010). It is still possible to do all the calculations for a t-test on the samples and declare significance when the resulting p-value is less than the traditional $\alpha = 0.05$. However, it must be remembered that the p-value is the probability of getting a particular difference in means assuming that the data are normally distributed and the samples being compared are drawn from the same normal distribution. If the samples are both drawn from a log-normal distribution instead, then the probability of such a difference in means might be more (or even less) likely than the α used to declare significance in the t-test.

To see what is really going on, researchers run simulations where the null hypothesis is known to hold. So for our example, random samples of particular sizes are drawn from a heavy-tailed distribution and compared using a t-test. Because they are from the same distribution, the null hypothesis is known to hold but of course chance variation tells us that the means will be different and sometimes will be significantly different, according to a t-test, purely by chance. Suppose we fix the level of significance at $\alpha = 0.05$ and do 10,000 such simulated t-tests, then if the t-test is robust we would expect to declare significance roughly 500 times ($500/10,000 = 0.05$). However, in practice, with heavy-tailed distributions, you see a lot more significant results, about 1,500 results with samples of size 40 (Wilcox, 2010, p. 73). Thus, you are more likely to make a Type I error than you might think and this would lead us to conclude that a t-test is not robust in this situation.

If a test is not robust to Type I errors then it is not going to be robust to Type II errors either, because β depends on α. However, there are many tests for which Type I errors are held at α over a useful range of distributions, in particular non-parametric tests tend to have good Type I error behaviour over a wide range of distributions. The question then becomes: to what extent is the test able to reliably see effects? That is, what is the power of the test?

There is no easy way to answer this because of the very situated nature of power in the design of the study where it is used. What is done, therefore, is to run another large set of simulations to see what differences a test sees when there are real differences to be seen. Again using a t-test as an example, to see what happens when homogeneity of variance is violated, particular sizes of samples are taken from two normal distributions whose variances (standard deviations) differ substantially but whose means also differ, to see if the t-test can still detect the difference in means despite the violated assumption. If, over 10,000 simulations, the t-test finds a significant result 4,000 times then the power is about 40%. This may be good or bad, depending on just how different the variances are and what the difference in means actually is. However, it serves as a useful basis for comparison with other tests, say a Mann–Whitney test. If in these circumstance the Mann–Whitney finds significance a lot more, say 4,500 times, then it would seem that in this situation the Mann–Whitney is the more powerful test.

Of course, this is just one comparison of two tests. There may be many more tests to compare, distributions to consider, assumptions to violate and effect sizes to manipulate. At the end of the day, there is a limitless number of distributions from which samples in real experiments may be taken. It is still impossible to know which test is actually best in which situation, but what can emerge is a picture of how tests vary in practical situations and whether one particular test tends to be more robust than another. That is, these sorts of analysis allow us to see when a test keeps the Type I error rate close to the nominal α value and, at the same time, provide better power than other related tests.

There are examples of exactly this sort of analysis and discussion in various chapters: Chapter 11 for t-tests, Chapter 10 for non-parametric tests, Chapter 12 for ANOVA and Chapter 18 for questionnaire and Likert item data. In particular, it is used as the main way in which claims of the robustness (or otherwise) of different tests are evaluated.

9.5   Summary

Power is a useful concept, both for helping to think about the quality of a specific study and also for assessing the quality, in particular the robustness, of different tests. However, it must be emphasised that power is at best hypothetical. For any particular study that is attempting to find out new effects, it is not possible to know what is real, what should happen, and therefore nor is it possible to know what the power of the study is. Similarly, though Type I error rates and power behaviours may indicate how good a test is, it is also not possible to know if any one test is definitively the best one for the study in hand.

Nonetheless, power helps us to think clearly and to make explicit some of the factors that might be in play. Even though power is deeply rooted in the NHST paradigm of statistical testing, it provides useful tools to help us think about both the design of studies and the quality of the tests. In this way, we can look beyond simply deciding whether a result is significant or not.

References

Cairns, Paul (2016). 'Experimental methods in human-computer interaction'. In: *The Encyclopedia of Human-Computer Interaction*. Ed. by Mads Soegaard and Rikke Friis Dam. 2nd. Interaction Design Foundation. Chap. 34.

Cohen, Jacob (1992). 'A power primer'. *Psychological Bulletin* 112.1, pp. 155–159.

Hoenig, John M. and Dennis M. Heisey (2001). 'The abuse of power: The pervasive fallacy of power calculations for data analysis'. *The American Statistician* 55.1, pp. 19–24.

Hornbæk, Kasper, Søren S. Sander, Javier Andrés Bargas-Avila and Jakob Grue Simonsen (2014). 'Is once enough?: On the extent and content of replications in human-computer interaction'. In: *Proceedings of the SIGCHI Conference on Human Factors in Computing Systems*. ACM, pp. 3523–3532.

Neyman, Jerzy and Egon S. Pearson (1933). 'On the problem of the most efficient tests of statistical hypotheses'. In: *Phil. Trans. of the Royal Society of London (A)* 231, pp. 289–337.

Robertson, Judy and Maurits Kaptein (2016). 'Improving statistical practice in HCI'. In: *Modern Statistical Methods for HCI*. Springer, pp. 331–348.

Simmons, Joseph P., Leif D Nelson and Uri Simonsohn (2011). 'False-positive psychology: Undisclosed flexibility in data collection and analysis allows presenting anything as significant'. *Psychological Science* 22.11, pp. 1359–1366.

Wilcox, Rand R. (2010). *Fundamentals of Modern Statistical Methods: Substantially Improving Power and Accuracy*. Springer Science & Business Media.

Yatani, Koji (2016). 'Effect sizes and power analysis in HCI'. In: *Modern Statistical Methods for HCI*. Springer, pp. 87–110.

Zendle, David, Paul Cairns and Daniel Kudenko (2018). 'No priming in video games'. In: *Computers in Human Behavior* 78, pp. 113–125.

Zimmerman, Donald W. (1994). 'A note on the influence of out'liers on parametric and nonparametric tests'. *Journal of General Psychology* 121.4, pp. 391–401.

CHAPTER 10

Using Non-Parametric Tests

Questions I am asked:

▷ Should I do a Mann–Whitney test or a *t*-test?
▷ Aren't non-parametric tests less powerful than parametric tests?
▷ Are non-parametric tests really more robust than parametric tests?
▷ How should I report a non-parametric test?

The term 'non-parametric test' is something of an odd term. Parametric tests typically refer to tests based on the assumption of normality. But even from the earliest studies into the bell-shaped curve, it has been known that, normally, data were not, well, …normal (Wilcox, 2010). Real data come in all sorts of shapes. 'Parametric-' or 'normal-data' was just one of those shapes but nonetheless the dominance of the normal distribution led to lumping all the alternatives together into a single class of non-parametric data. However, though there are a large number of tests designed to deal with non-parametric data of all varieties, the term 'non-parametric tests' is typically used to refer to those tests that involve ranking the data before doing any analysis. The classic non-parametric tests are the Wilcoxon, the Mann–Whitney, the Kruskal–Wallis and the Friedman tests (see, for example, Howell (2016)). It is these rank-based tests that are the concern in this chapter.

The increased relevance of non-parametric tests in recent times has grown in part from the recognition that the assumption of normality is neither reliably true nor safe to overlook (Chapters 7 and 11). In particular, rank-based tests are sometimes said to be distribution-free, that is, do not depend on any assumptions about the underlying distribution. However, that is not quite true. Rank-based tests are also sometimes said to be a test of the median rather than the mean, but that is not quite true either. The purpose of this chapter is to unpick some of the common claims about non-parametric tests and so help to inform both a sensible use of the tests, the use of newer but less well-known alternatives and a sound interpretation and reporting of their results.

10.1 The Mechanics of Ranks

I have to confess to a certain fondness for non-parametric tests. Anyone who has learned statistics formally is usually at some point asked to calculate test statistics by hand. With *t*-tests and ANOVA, this rapidly becomes some long, complicated arithmetic with squares and square roots flying around. With non-parametric tests, there is very little of that. Typically, the data are ranked in some way, which needs care but is very straightforward. And then ranks are summed in some way and, in some cases, like a Mann–Whitney test, that's it! The generated test statistic is then compared to a standard table to give a *p*-value and the calculation is over.

It also feels more transparent than parametric statistics. For example, consider a simple situation where we might use a Mann–Whitney test. Here, I follow the procedure in Howell (2016). Suppose I set up an experiment to see how people's level of spatial presence in a game, the sense of being physically located in a virtual environment (Lombard and Ditton, 1997), is affected by using a virtual reality (VR) headset, expecting more spatial presence with VR. Spatial presence is measured on a scale from 1 to 10 and is highly specific to the different contexts, so we cannot assume the data will come from a normal population (remember, normality is a property of the population, not of the sample, Chapter 7). And I am a very optimistic researcher so I collect the data from only ten participants, five in each condition, in the hope that that will be enough. The data might look like the following:[1]

	VR	No VR
	5	3
	8	5
	2	6
	7	1
	9	7
Mean	6.0	4.4
Median	6	5

---

[1] This example is deliberately underpowered just to make the example small.

The mean and median are higher for the VR condition, which is promising. Transforming the data to ranks from 1 to 10 gives:

	Rank VR	Rank No VR
	4.5	3
	9	4.5
	2	6.5
	6.5	1
	10	8
Sum	32	23

Remember that when two scores tie, they share the ranks for the number of places that they take up, so that 5 being both the 4th and 5th ranked scores gets ranked as 4.5 both times it occurs.

Here's where the transparency comes in. The ranks from 1 to 10 add up to 55 (you can use triangle numbers to work this out quickly). If the null hypothesis holds then the summed ranks of the two groups should be roughly equal and, moreover, should add up to 55, so a null result would give both summed ranks to be about 27.5. The question of significance in this example is whether 32 (or 23) is sufficiently far from 27.5 to be unlikely to have occurred by chance.[2] There is an intuitiveness about this that is immediate in contrast to the corresponding (and inappropriate) t-test, where we would need to look up a critical value of t for 8 degrees of freedom. It takes a lot longer to develop intuitions about t values.

The other common non-parametric tests work similarly. The Kruskal–Wallis test corresponds to the Mann–Whitney test when there are more than two conditions to compare. Thus, it is a non-parametric equivalent of a one-way, between-participants ANOVA. It uses a ranking procedure for converting raw scores to ranks that is identical to the Mann–Whitney. For within-participant designs, the Wilcoxon is used for two conditions and ranking is actually done on the difference between each participants' scores. And when there are more than two conditions in a within-participant design, the Friedman test is used. It has a quite different, but even simpler, ranking procedure where all of the scores for a participant are ranked just for that participant. In all cases, it is easy to translate from a null hypothesis to what the behaviour of the rankings should be like. So, if you are doing the calculations by hand, it can be fun to see how differences in data lead directly to significance. Of course, statistics packages have made such simple pleasures completely obsolete.

---

[2] Turns out $p = 0.293$ so not significant. Looks like I really was too underpowered!

10.2 Analysing Errors

To understand how to interpret non-parametric tests, it is first useful to examine their performance with regards to Type I and Type II errors to see how robust they are as tests (Chapter 9). The two types of error are unavoidable in any statistical analysis because of the inherent uncertainty and randomness in data, but good quality tests manage the errors in predictable and quantifiable ways.

10.2.1 Type I Errors

A Type I error, recall, occurs when a test gives a significant result when the null hypothesis is in fact correct. This can always happen by chance but, in the context of severe testing (Chapters 1 and 2), a significant result should reflect a result that is unlikely to have happened at the chosen level of significance and so indicate support for the idea being tested.

For non-parametric tests in the context of an experiment, the null hypothesis is that there is no difference between the conditions. In our example, the use of VR did not affect people on their rating of spatial presence, and people's experiences were broadly the same in both conditions within the parameters of the typical variations of the population in general. It is well known that non-parametric tests, by assuming no specific underlying distribution and reducing data to ranks, preserve the Type I error rate reliably at the nominated level of significance. Put another way, over a large number of repeated evaluations when the null hypothesis really does hold and α, the level of significance is 0.05, then non-parametric tests make a Type I error about 5% of the time.

It is because of this robustness with respect to Type I errors that non-parametric tests are considered distribution free. However, notice that the null hypothesis literally means the experimental manipulation has no effect whatsoever, not just on mean or median values.

However, the null hypothesis is not always quite so simple. Consider a usability test where two different systems are being evaluated for the time it takes people to do a task. There are likely to be substantial differences between the designs of the systems, and so it is not likely that the sets of people using the two different systems could be expected to behave like a single population. In this sense, the null hypothesis of the usability test is not that people are completely unaffected but instead that there is no meaningful difference. For instance, what is typically asked in this situation is whether 'on average' people are quicker with one system than another.

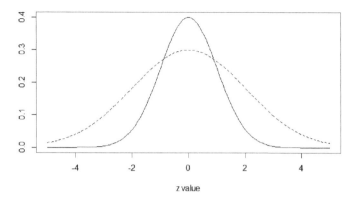

z value

Figure 10.1   Two distributions that differ only in variance but nonetheless affect the Type
I error rate of the Mann–Whitney test.

In more technical language, is there a change in location of the underlying
data due to which system is used? (Chapter 4)

In this scenario, a non-parametric test still seems perfectly appropriate.
However, Kasuya (2001) showed that for normal distributions where the
mean and median were the same, (no change in location), but the standard
deviations differed between the samples (as in Figure 10.1), the Mann–
Whitney test had an increased rate of Type I errors. That is, the result was
significant more often than the nominal $\alpha = 0.05$ would suggest. Even
quite modest differences in standard deviation, say one group having twice
the standard deviation of the other, could result in the actual Type I error
rate of being above 0.1, particularly when sample sizes between the two
conditions were different. That is, if a significant result was interpreted as a
change in location then in fact it is more realistic to assume $\alpha = 0.1$, which
is probably not what researchers would want to do.

Similar results were shown in a wider variety of distributions by Zimmerman (1998). Moreover, this is not a problem of sample size because Pratt
(1964) showed that this inflation happens asymptotically, which is to say for
arbitrarily large sample sizes. This was even in what might be considered
ideal situations where samples were the same size and the distributions were
both normal.

Arguably, the rate of getting significance *ought* go up as the null hypothesis does not in fact hold: the underlying distributions are actually different
in their standard deviation. But if the interpretation in our usability test is
that there is change in location (mean or median) between the two groups,
then this is incorrect. There is a difference in the distributions of the two

groups but it is due to variance not averages. A difference in variance is unlikely to be of much relevance to the usability of the systems. Even if it were, the Mann–Whitney test would not be the test to use because, even though it is affected by differences in variances, it is not designed to test differences in variance. It is possible to have differences in variances that a Mann–Whitney could not guarantee to find even with enormous sample sizes. In technical terms, the Mann–Whitney is not a consistent test against finding differences in variances (Zaremba, 1962).

Differences in variance can be examined in the selected samples and then caution used in interpreting the tests. However, what becomes a real problem is the situation when differences in variances get mixed up with small effects. The difference in variances increases the chance appearance of significance, so a small difference in mean or median time on task that is not necessarily very meaningful could be inflated to the level of significance by a difference in variances. This could lead to misleading conclusions about the effectiveness of one system over the other.

10.2.2   Type II Errors

Type II errors occur when there is systematic difference between groups of people or groups of data but a statistical test fails to give significance. It is well-established folklore that non-parametric tests are less powerful than parametric tests and so more likely to make a Type II error. However, this is *only* when the underlying data are ideal for parametric tests. Even then, a small increase in sample size of about 10% restores the power of the non-parametric to the same level as the parametric test. What really matters is that in situations where a *t*-test or ANOVA can be very unreliable because normality is simply not present, a non-parametric test can outperform parametric tests by far.

One particular case where this is most easily seen is with outliers. The ranking process reduces the impact of any outliers to be within the range of ranks possible for the sample size. Thus, unlike a *t*-test where an outlier can have an arbitrarily large influence on the outcome (Chapter 8), non-parametric tests are not strongly affected by outliers. Only when the likely proportion of outliers in a sample becomes appreciable is there a reduction in the power of a test like the Mann–Whitney (Zimmerman, 1994). At that point though, it is an interesting question as to when a large proportion of outliers becomes less about outliers and more a characteristic of the data. See Chapter 8 for a more detailed discussion around this.

Another, commonly overlooked, feature of rank-based tests is that they are affected by differences in shapes of the distributions for different experimental conditions. This is an extension of the situation leading to Type I errors, described above, where a difference in variance of the distributions led to increased likelihood of getting significance when the means and medians were the same. By contrast though, in this situation, a difference in the shape of distributions leads to reduced power, for example when one sample is from a normal distribution and another from a long-tailed distribution (Wilcox, 2010, p. 221). In such a situation, even though there can be differences in distributions that the Mann–Whitney could in principle detect, it is not able to do so reliably enough to give significance at even modest sample sizes

This seems very counter-intuitive to what is usually talked about for non-parametric tests, though it does highlight that non-parametric tests are not distribution free. Another way to see why non-parametric tests are not distribution free comes from the work of Conover and Iman (1981). They showed that some non-parametric tests are equivalent to conducting a parametric test on the same data after they have been converted to ranks. In particular, conducting a Mann–Whitney test on the first table of raw scores in Section 10.1 is exactly equivalent to conducting a t-test on the rank scores of the second table in Section 10.1. Looked at this way, a Mann–Whitney (and other rank-based tests) are susceptible to all the same problems that can reduce the power of parametric tests, including unequal variances, skew and so on. The non-parametric tests tend to be *less* affected than their parametric equivalents but they are nevertheless affected. It is also for this reason that procedures that make parametric tests more robust (Chapter 11), also make non-parametric tests (even) more robust (Zimmerman, 1995).

10.3   Practical Use

Given all the problems and flaws in non-parametric tests already detailed, it may seem that they are just not worth the bother. Might as well use a dodgy parametric statistic as a dodgy non-parametric one! However, this overlooks the practical situations where we use tests and that some theoretical situations are not relevant to our particular goals.

It should first be noted that there do not seem to be any particular concerns for the behaviour of the Wilcoxon and Friedman tests. This seems to be for two separate reasons. First, because the Wilcoxon works only with a single set of difference scores, much like a one-sample t-test, all the

problems of differences in variance and shape that led to inflation in both
Type I and Type II error rates in the Mann–Whitney are not applicable.
Second, the Friedman test actually functions like a generalisation of the
Sign test, not the Wilcoxon test, for multiple conditions (Zimmerman and
Zumbo, 1993). The Sign test, and therefore the Friedman test, are very
simple tests that do not rely on the magnitude of scores at all, simply the
direction of changes in score between conditions of an experiment. Thus,
the Friedman is not susceptible to problems of variances either.

This leaves the Mann–Whitney and Kruskal–Wallis tests. What they
really test is whether there is one condition or group in a study where the
scores of that group are likely to be larger than the scores of another group.
That is, they are tests of dominance, not tests of location (Chapter 4). More
formally, they look for a situation where there are two groups from the
study, say A and B, and the scores, X_A, from A are more likely to be bigger
than the scores, X_B, from B. In terms of probabilities:

$$P(X_A > X_B) > 0.5$$

That is, the tests look to see if group A dominates group B. However,
the snag is that these tests cannot necessarily reliably detect this when
variances are unequal (Zaremba, 1962). This is particularly the case when
one distribution 'spreads out' over the other, as in Figure 10.2.

Fortunately, in the context of controlled experiments, this situation is
unlikely. What is more likely is that a small change is made to experimental
conditions and that leads to a movement of scores one way or another
between the conditions but without drastically altering the shape of the

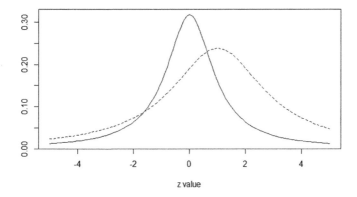

Figure 10.2   Two distributions that differ in dominance but where one also spreads out
across the other making a Mann–Whitney test low in power.

distribution. And even if it does affect the shape, it does so in a way that does not drastically increase spread and hence lead to the poor performance of non-parametric tests. In particular, if the shape stays roughly the same, non-parametric tests give good performance for detecting differences in dominance in the face of arbitrary distributions and in particular with robustness against outliers.

In the context of usability tests though, it may be a different story entirely. The general behaviour of people as seen through variables like number of errors or time on task could change dramatically when systems change substantially. Thus, the shapes of the underlying distributions being studied may no longer be of the sort that non-parametric tests can cope with. At this point, it is time to retreat from an obsession with significance and look at the distribution of the data collected from each condition alongside the qualitative differences in people's behaviour. It may be that in this situation, statistical tests are just not as informative as more careful analysis of what people actually get up to when using the different systems.

How can we know when a usability test, or indeed any other study, might lead to a violation of the assumptions underlying non-parametric tests? As with any assumption about the populations, we can't! (see Chapter 7 for a discussion of this problem in relation to normality.) However, experience of similar studies, knowledge of the measures used in the studies and care in analysis will all help guard against problems.

Furthermore, there are two further non-parametric tests that might help. The Cliff test and the Brunner–Munzel test are both tests of the dominance of one group over another group (Wilcox, 2017). They have been shown to be robust to differences in variance of the underlying distributions, even when sample sizes are small (ten participants in each group) and so offer a useful alternative to the traditional Mann–Whitney test (Neuhäuser, Lösch and Jöckel, 2007). Sadly, these tests lack the transparency of rank-based tests, but that is an aesthetic problem, not a practical one. For these tests, the actual statistic produced is $\hat{P}$, the measure of dominance, and so directly related to the precise hypothesis that they test.

10.4   Reporting Non-Parametric Tests

When reporting *parametric* tests, the standard expectation is to report means and standard deviations for every separate group of data that is compared. Each parametric test typically has a specific, well-established measure of effect size associated with it as well. Of course, these are not meant to be valid measures for non-parametric data so it leaves a question of what should be reported.

One way to side-step this is to present a boxplot! It gives a lot of useful data that capture characteristics of a sample, whether it is expected to be parametric or not. However, it does seem conventions and expectations have not caught up with non-parametric tests. I find that people still expect to see means and standard deviations. I do not see the problem with reporting these either (as long as there is a boxplot too), particularly as medians can sometimes be unrepresentative of a sample (Wilcox, 2010, p. 43). There could be an argument for replacing traditional standard deviations with robust estimators of standard deviation based on interquartile ranges or median absolute deviations (MAD) as used to classify outliers (Chapter 8). These seem like useful alternatives for non-parametric situations.

In addition to these traditional descriptives, probably the best further descriptive is the measure of dominance, $\hat{P}$, that is the statistic that evaluates $P(X_A > X_B)$ (see Chapter 4). This is after all what these rank-based tests are actually examining. Even if the p-value might be called into question, the dominance of one group over another can help to see whether things are moving in the direction as predicted for an experiment or usability study.

10.5   Summary

Rank-based non-parametric tests do offer some robustness over their parametric counterparts. All of the tests are more robust to problems with outliers because of the ranking process. Also the Wilcoxon and Friedman tests seem very reliably robust in all practical circumstances. The Mann–Whitney and Kruskal–Wallis tests are less robust and, in particular, susceptible to changes in variances between samples, in which case they produce both an increased chance of Type I errors and Type II errors. Researchers should be cautious about using these tests without thinking about the changes in distributions that they are looking for and when in doubt should opt to use the Brunner–Munzel or Cliff tests in place of the Mann–Whitney test.

References

Conover, William J. and Ronald L. Iman (1981). 'Rank transformations as a bridge between parametric and nonparametric statistics'. *The American Statistician* 35.3, pp. 124–129.

Howell, David C. (2016). *Fundamental Statistics for the Behavioral Sciences*. Nelson Education.

Kasuya, Eiiti (2001). 'Mann–Whitney U test when variances are unequal'. *Animal Behaviour* 61.6, pp. 1247–1249.

Lombard, Matthew and Theresa Ditton (1997). 'At the heart of it all: The concept of presence'. *Journal of Computer-Mediated Communication* 3.2. Available at: https://onlinelibrary.wiley.com/doi/abs/10.1111/j.1083-6101.1997.tb00072.x [Accessed October 1, 2018].

Neuhäuser, Markus, Christian Lösch and Karl-Heinz Jöckel (2007). 'The Chen–Luo test in case of heteroscedasticity'. *Computational Statistics & Data Analysis* 51.10, pp. 5055–5060.

Pratt, John W. (1964). 'Robustness of some procedures for the two-sample location problem'. In: *Journal of the American Statistical Association* 59.307, pp. 665–680.

Wilcox, Rand R. (2010). *Fundamentals of Modern Statistical Methods: Substantially Improving Power and Accuracy*. Springer Science & Business Media.
(2017). *Introduction to Robust Estimation and Hypothesis Testing*. 4th edn. Academic Press.

Zaremba, S. K. (1962). 'A generalization of Wilcoxon's test'. *Monatshefte für Mathematik* 66.4, pp. 359–370.

Zimmerman, Donald W. (1994). 'A note on the influence of outliers on parametric and nonparametric tests'. *Journal of General Psychology* 121.4, pp. 391–401.
(1995). 'Increasing the power of nonparametric tests by detecting and down-weighting outliers'. *The Journal of Experimental Education* 64.1, pp. 71–78.
(1998). 'Invalidation of parametric and nonparametric statistical tests by concurrent violation of two assumptions'. *The Journal of Experimental Education* 67.1, pp. 55–68.

Zimmerman, Donald W. and Bruno D. Zumbo (1993). 'Relative power of the Wilcoxon test, the Friedman test, and repeated-measures ANOVA on ranks'. *The Journal of Experimental Education* 62.1, pp. 75–86.

CHAPTER 11

A Robust t-*Test*

Questions I am asked:

▷ Aren't *t*-tests robust so I don't need to worry about assumptions of
normality?

▷ My data are not normal, but can I still use a *t*-test?

▷ My data are not normal, so should I transform the data to make them
normal?

▷ Should I test for assumptions before I use a *t*-test?

▷ Okay, so I can't use a *t*-test, but what can I use?

When it comes to thinking statistically, the mean of a population or a
sample is a familiar, immediate and useful concept. It is the arithmetic
average of a sample and this makes it an intuitive statistic in lots of
situations. And where the population distribution of a measure is believed
to be symmetric, it is easy to show that the mean is also the median of the
distribution and the point of symmetry (Rosenberger and Gasko, 1983).
If, further, the population is normally distributed then the mean is both a
key parameter in completely describing the population and has even more
useful properties (see Chapter 7).

Given the dominance of means, it is therefore not surprising that the
t-test (or more precisely, the family of tests collectively called the *t*-test),
which is able to test for differences in means, has been a key test in
the history of statistics. Strictly speaking, the *t*-test relies on assumptions
about the underlying distribution, but there is also historic evidence
that parametric tests, including the *t*-test, are robust to deviations from
assumptions including unequal variances, particularly when sample sizes
are equal (Box, 1954; Sawilowsky and Blair, 1992). All in all, the *t*-test seems
rightly to hold its place in the analysis toolkit of modern researchers.

However, there is growing evidence that the *t*-test is not as robust as we might think nor as we might like it to be. While the older research about robustness was done well for its time, newer results are showing that those analyses were somewhat limited. More realistic situations for testing the robustness of *t*-tests suggest that they can fail to hold both Type I and Type II error rates, making them of dubious value, particularly where little is known of the underlying distributions of the data (Wilcox, 1998).

If *t*-tests are not reliable, what then is the best way to compare means? It is possible to move to non-parametric tests but, as discussed in Chapter 10, those tests are best understood to test for dominance, the propensity for one set of scores to be bigger than another set. This is one sort of useful outcome but, where there is a need to quantify the difference, say for instance to see if a system really does make it measurably quicker to perform a task, then this is not enough. Stepping back from a fixation on means, what is really needed is to see changes in the location of a distribution (Chapter 4). Means are the right measure of location in normal distributions but not necessarily in other distributions, particular when distributions might differ in variance and skew or have outliers.

The purpose, then, of this chapter is first to explain a little both of how a *t*-test works and also why it may fail to work in realistic situations. I will then discuss the use of robust estimators of location that could replace means and the tests that could be used to assess differences in location between groups. Specifically, trimmed means, Winsorization and M-estimators are considered as effective alternatives to means and how tests that use these measures offer robust alternatives to *t*-tests.

11.1   A Traditional *t*-Test

Suppose we run a study looking at improving the use of a library search system. There is the traditional search-box version, S, and the new version, F, which suggests filters based on the queries users enter. We collect time to complete the task, T, of finding a useful statistics book. This is of course open ended, to be realistic, and it is up to the participants to decide when they've stopped. In this study, two sets of times are therefore collected, T_S and T_F, from ten users of each system and ideally we are hoping to see a meaningful reduction in task time using the filter system, F.

This is a typical sort of situation in which a *t*-test is used in HCI. The formula for the *t*-value for the study is:

$$t(18) = \frac{\overline{T_F} - \overline{T_S}}{s_e} \tag{11.1}$$

That is, the *t*-value for 18 degrees of freedom (based on 20 participants in 2 groups) is the difference in the means for each set of times divided by an estimate of the standard error, s_e. The standard error is the standard deviation of the sampling distribution for the means. Roughly, but not precisely, it is the average standard deviation of the two sets of times divided by the square root of the number of participants in a group.

The *t*-test looks at the *t*-value and if it is larger than might be expected by chance alone, the null hypothesis, then the *p*-value is small and the difference in means is significant.

A *t*-test makes intuitive sense. The question it addresses is whether there is a big difference in the mean times from the two groups of participants. What does 'big' mean? Well, thanks to the central limit theorem (Chapter 7) the two means should come from a normal distribution and so it is a question of how big the difference is in relation to the standard deviation that might be seen for means of this sort. That is, the difference in means is big if the ratio of the difference in means to the standard error is big, which is precisely what the *t*-value gives us.

However, underlying the *t*-test are three key assumptions:

1. Independence of observations
2. Normality
3. Homoscedasticity

This first assumption means that what you measure about one thing does not influence what you measure about another. In the example, this would be that the time a person takes to find a useful statistics book is independent of the time somebody else took. This is almost always accounted for in the experimental design, which should prevent participants from influencing each other (such as one participant telling another which book they found). However, independence of observations, though typically just assumed, is worth pointing out because it becomes relevant to robust alternatives to the *t*-test.

Normality is the assumption that both sets of times come from a normal distribution. However, as discussed in Chapter 7, that assumption is both hard to justify and moreover hard to ignore. Skew, different distributions and outliers can all cause the assumption of normality to be inappropriate and undermine the ability of the *t*-test to give a clear result.

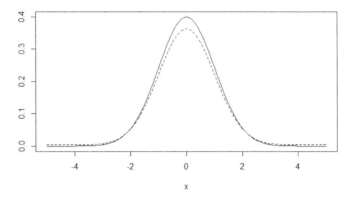

Figure 11.1   The probability distribution curve of a normal distribution with standard
deviation of 1 and a mixed-normal curve with a standard deviation of 3.3.

Possibly the most convincing example of this is when data come from a mixed-normal distribution. This theoretical distribution is a combination of two normal distributions such that most of the time data come from one distribution, N_1, with say a small standard deviation but with a small probability, q, come from a distribution with a much larger standard deviation, N_2. The means of the two distributions are the same and so result in the same mean for the mixed distribution. Such distributions can be indistinguishable by eye from a normal distribution and seem only marginally different from the unmixed N_1, see Figure 11.1, but in fact the standard deviation of the mixed distribution is a combination of the standard deviations of both distributions so that, even though q is small, the standard deviation of the mixed distribution can be a lot higher than N_1 would suggest.

The result of a high standard deviation is that the t-test on data from the mixed normal distribution gives much smaller t-values and so is unlikely to detect effects even if there are useful differences in location to be found. That is, there is an increase in Type II errors (Chapter 9).

This may seem like a theoretical trick to undermine t-tests but in the context of the example study, it would be realistic to assume that there is the general library user population but also some library users who are whizzes at statistics, and so are either much quicker to do the task, as quicker to identify suitable books or much more deliberate in their choices to make sure they do not skip over non-obvious useful results. On *average* they look the same but they have a much higher standard deviation in their times. Such a mix of two types of participants would be the basis for a mixed-normal distribution and any *t*-tests would be unreliable.

The final assumption of homoscedasticity is that the standard deviations, or more strictly speaking the variances, of the two groups are the same. A lack of homoscedasticity is called heteroscedasticity. The idea is that in situations of comparing two groups, only the mean changes and the variance is unaffected by a change in means. In practice this is rarely the case because if an experimental situation is having an effect, it tends to cause bigger effects in some participants and smaller effects in others, with the result that even if all the effects lead to an increase in means, there is also a related increase in variance.

Early researchers on the robustness of the *t*-test were concerned about what would happen if there were significant departures from homoscedasticity and in particular what would happen to the Type I error rate (Chapter 9). Provided the sample sizes in the two groups are the same, the Type I error rate is pretty reliable both theoretically as samples get arbitrarily large (Pratt, 1964) and practically for smaller samples (Box, 1954; Ramsey, 1980) regardless of the degree of heteroscedasticity. This is, of course, provided distributions are normal. And even where distributions are not normal, provided the null hypothesis is that nothing at all changes in the distributions, then the *t*-test still gives reasonably good control of Type I errors with realistic, non-normal distributions and equal sample sizes (Sawilowsky and Blair, 1992).

It is these results that have led to the claims that *t*-tests are robust to the violation of assumptions. However, as soon as sample sizes are unequal, there is no guarantee how Type I errors will behave, even for very large samples from normal distributions (Pratt, 1964; Ramsey, 1980). Moreover, when data are not normal, like the mixed-normal distribution, or, worse, when there are both non-normality and heteroscedasticity, then Type I error rates can be out of control and power is lost as well (Algina, Oshima and Lin, 1994). The comforting results of early researchers simply did not look at these more complicated yet realistic situations.

Overall then, the assumptions of normality and homoscedasticity are strong but do not reliably hold in realistic data. This threatens the value of *t*-tests and cannot be ignored. At the same time, a comparison of means provides an intuitive and meaningful way of describing what might be happening in a study. What is needed, then, is a way to gather information about the change in location of data due to experimental conditions without hitting the problems of not knowing in advance the shape of the underlying distribution of the data.

11.2   Simple Solutions?

One approach commonly seen is to test first for violations of the assumptions. There are issues with this.

The Levene test for equality of variances works well in checking for homoscedasticity provided distributions are symmetric. When they are skewed, the Type I error rate is too high, which suggests heteroscedasticity is significant even when it is not (Brown and Forsythe, 1974). Similarly, tests for normality like the Kolmogorov–Smirnov and Shapiro–Wilk test are not powerful enough to detect deviations from normality on small samples, as discussed in Chapter 7. Furthermore, detected deviations from normality may not be relevant to what makes t-tests unreliable.

Testing for assumptions to see if a t-test is suitable is pretty much a futile exercise and not recommended (Erceg-Hurn and Mirosevich, 2008).

Another commonly seen solution is to transform the data. That is, before analysing the data, a function is applied to each item of data, such as taking the logarithm or square root. The idea is that such transformations adjust data so that they are more likely to meet the assumptions by reducing skew or the effect of outliers (Emerson and Stoto, 1983). However, it has the unfortunate effect of changing what is being analysed. For example, if we were to take logarithms of our study data, the t-test would actually end up evaluating the ratio of the geometric means of the original data in relation to the variation in the *ratio* of each data point to the geometric mean. I have no idea how to interpret that! Of all possible standard transformations that are proposed, only one transformation makes intuitive sense to me and that is taking reciprocals of timings, (T goes to $\frac{1}{T}$), so that effectively speed is analysed instead of time.

Even though there is some evidence that transformations can help sometimes, there is also evidence that they can reduce the power of a test without solving any of the problems. There is, of course, debate but Grissom (2000) provides a useful review. On the whole, it seems that transformations do not really overcome violations of assumptions. More problematic for interpretation, a transformation muddies the fact that normally the t-test is linked to means and a change of location. The result of a t-test on transformed data is about the transformed means, which may not have any intuitive link to the location of the underlying data.

Though transformations of data are occasionally seen in HCI, one correction for heteroscedasticity is not commonly seen but is in fact much more reliable. This is a relatively old variation on a t-test due to Welch (1947). What Welch showed was that if the expression for the t-distribution underlying the t-test is analysed the right way, heteroscedasticity can be compensated for by adjusting the degrees of freedom, which are used to generate the *p*-value from the *t*-value. Interestingly, this is not a correction like a transformation, to be used when things are not right with the data, but is in fact a general improvement to the *t*-test. This makes the Welch test rather appealing because everything works the same as a *t*-test until the last step of generating the *p*-value.

It is hard to know why the Welch test has not been more widely promoted in textbooks, as it genuinely relies on fewer assumptions than the *t*-test. It is more complicated to calculate, so perhaps when people believed (a) everything was normally distributed, (b) *t*-tests are robust anyway and (c) computers were not around to help, it may have been seen as simply not worth the effort. Of course, the Welch test will not solve problems due to non-normality but it will not make them worse.

11.3 Location, Location, Location

A feature of the mean of a distribution that makes it so useful is that it is a measure of location. That is, it indicates where the distribution is, so that if a distribution moves wholesale due to an experimental manipulation, as in Figure 4.1a, the means change by the same amount.

Thinking about this though, any number that is primarily a function of the shape of the distribution will work well as a measure of location of the distribution. For example, the lowest value of the distribution (if it has one), the median, or either quartile or indeed any quantile (a value that splits the distribution into a fixed proportion). There are a whole host of possible values.

The mean also (usually) has the useful property that it is somewhat representative of the distribution as a whole. When distributions are symmetric, it is literally the middle of the distribution. When the distribution is not symmetric though, this is not the case. Usually, good measures of location are also representative, that is meaningful to the data as a whole, and might even be plausible as data values.

The problem is not finding a measure of location but finding one that can be robustly estimated from samples. The mean is not robust because, as discussed in Chapter 8, the mean of a sample is easily influenced by outliers and so can be unrepresentative of the distribution as whole.

The median, therefore, looks like a good candidate as a robust measure because it is not affected by outliers or even a large proportion of outliers. But estimating the median robustly can fail for other reasons. Consider the Likert scale distribution represented in Figure 11.2.

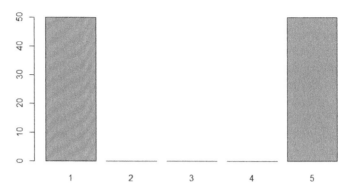

Figure 11.2   A polarised distribution on a Likert scale: 100 people are equally divided
between those who strongly agree and those who strongly disagree.

The median for this distribution, by convention, is 3. But the median of a sample is either 1, 3 or 5 and only ever 3 (the true median) if the sample is both of even size and there are exactly as many 1s in the sample as 5s. Thus, a sample whose median is the same as the population is actually quite unlikely to occur for large samples! So medians have the undesirable property that, even as samples get larger, the median of the sample cannot be guaranteed to converge on the median of the underlying distribution. Interestingly, in this example, the mean of samples would actually better approximate the distribution median as sample sizes grow.

Admittedly this is an extreme example and there is not much change of location that is even possible for this distribution, but it highlights that medians are not necessarily well-behaved estimators of the location of distribution, particularly when there is the possibility of tied scores in the data.

Statisticians have understandably thought a lot about this and developed measures of location that not only share the representativeness of means and medians but can also be robustly estimated from samples. This makes them a good foundation for effective alternatives to the *t*-test. Wilcox (2017, chap 1–3) gives a thorough account of the principles of robust estimation. There are three main measures that emerge from this research: trimmed means, Winsorized means and M-estimators.

11.4   Trimmed and Winsorized Means

A trimmed mean is the mean of a sample having trimmed out (thrown away!) a certain proportion of the data. If this is the first time you have come across trimming and you have run studies in the past, this may seem like madness. You go to all the effort of collecting data only to throw a big chunk of it away! Be reassured: this is not quite what is happening.

Suppose our study gathers times, T_S, which have been conveniently ordered, as follows:

T_S (s)	45 89 121 132 133 142 147 155 178 191 257 603	Mean (s) 183.8

Even a quick glance will tell you that there are some outlying values! The mean of the sample is 183.8s and the outlier of 603 is more than three times the mean. Moreover, the mean is in the top third of the data!

A γ-trimmed mean is one in which the top and bottom γ of the ordered values are omitted from the calculation. So a 10%-trimmed mean would neglect the 1.2 lowest and highest values. By convention, the fraction is rounded down, so only the lowest and highest values are neglected. Some trimmed means are shown:

γ		Mean (s)
10%	89 121 132 133 142 147 155 178 191 257	154.5
20%	121 132 133 142 147 155 178 191	149.9
25%	132 133 142 147 155 178	147.8
40%	133 142 147 155	144.3

What should immediately be seen is that trimming brings the mean closer to the median (144.5) of the data. In fact, a median would be a 50%-trimmed mean. (Well, almost because a true 50% trim would delete 100% of the values!) This helps to make sense of trimmed means. The data are not really being thrown away: if they had not been collected it would not be possible to know which data items represented the middle of the sample. Or alternatively, if you can be happy that a median is a useful measure of location, then a trimmed mean is a less extreme trim than a median.

Empirical evaluation of trimming has arrived at recommendation that the best amount of trimming is when $\gamma = 20\%$ (Hill and Dixon, 1982; Wilcox, 2017). This gives good results when no trimming is actually necessary and provides a robust estimate of location across across a range of underlying distributions when means are not adequate.

If trimming still seems rather casual about hard-won data, a less extreme process is Winsorization. Acknowledging that outliers are a problem but nonetheless representative (provided the data are sound, as discussed in Chapter 8), Winsorization aims to restrict the influence of outlying points.

To do this, instead of trimming a proportion, γ, of the data points, the points that would be trimmed are forced to have the value of the nearest acceptable data point. Our data would then become:

γ												Mean (s)
10%	89	89	121	132	133	142	147	155	178	191	257	257	157.6
20%	121	121	121	132	133	142	147	155	178	191	191	191	151.9
25%	132	132	132	132	133	142	147	155	178	178	178	178	151.4
40%	133	133	133	133	133	142	147	155	155	155	155	155	144.1

Like the trimmed mean, Winsorization brings the mean closer to the median. Whereas the trimmed mean neglects extreme values, Winsorization represents them all but at a less extreme value. This means that if there is considerable skew, say positive skew, in the distribution, the upper value that extreme values are Winsorized to could be further from the median than the lower value, and this would pull the Winsorized mean more towards that upper value. This would appropriately reflect in the mean the fact that there is skew in the data. In practice, though, it seems there is little to choose between trimming or Winsorization (Dixon, 1980).

11.5   M-Estimators

The traditional mean has a really useful mathematical property related to minimisation. If you think of standard deviation as the average distance of every point from the mean, it is possible to think of other deviations that are the analogous average distance away from some point other than the mean. For instance, we might have a median deviation, which would be the standard deviation calculated only relative to the median rather than the mean. The mathematical advantage of the mean is that it gives the least deviation of all such possible deviations and hence is why the standard deviation has become the standard.

To express this argument mathematically for our example, assume m is some fixed value, the variance about m of a sample of 10 task times T_S, given by

$$\text{var}(T, m) = \frac{\sum_{i=1}^{10}(T_i - m)^2}{9} \tag{11.2}$$

The mean is the special value of m that makes var(T, m) as small as possible for the data collected and so equivalently makes the standard deviation as small as possible.

More abstractly, if we define a function $s(x, m) = (x - m)^2$, the mean minimises $\sum s(T_i, m)$. A different function, $a(x, m) = |x - m|$, is minimised in the same way by the median. Looked at through these descriptions, useful measures of location can be generalised to be special values that minimise certain sorts of variation in the data.

M-estimators are just such general measures of location but where the function to be minimised, sometimes denoted ξ, has good properties, not just in terms of representing the distribution but also being robustly estimated by samples when the underlying distributions are not well-behaved. There are some good introductions to M-estimators that describe the different functions used in more detail, such as Goodall (1983) and Wilcox (2017).

The basic idea of M-estimators is that around the middle of a sample, ξ is like the usual estimate of variance (and hence a mean value would be appropriate) but at the edges of a sample, where there might be outliers or heavy-tails, ξ reduces the influence of these values but still factors them in. In this sense, M-estimators are also a generalisation of trimming and Winsorization.

The problem with calculating M-estimators is that, unlike means and medians, there is not some simple way to know advance how to explicitly calculate the M-estimator of a sample for a particular function, ξ. Thus, there needs to be an iteration where a first guess of the M-estimator is made, say using the median, and this is iteratively refined to minimise ξ for the data.

11.6   Back to *t*-Tests

Having devised these robust measures of location, it is not enough simply to stick them back into a *t*-test. This is for two reasons. First, even if the measures of location are robust, the usual standard deviation is not, and this could undo any of the value of working with these other measures. Second, and more importantly, because different items of data are included in the analysis depending on their order in the sample, the individual items of data that end up being used in the test are not independent of each other. This violates the first assumption for a *t*-test, namely the independence of observations. Avoiding one set of violations of assumptions has replaced it with a different one.

Fortunately, statisticians are well aware of this. Yuen (1974) developed a variant of the *t*-test that works specifically for trimmed means. Roughly, the calculation is the same as that in Equation 11.1: the difference of trimmed means is compared to an estimate of the standard error to give a traditional *t*-value. The difference from a normal *t*-test is that the standard error is based on the standard deviation of the *Winsorized* version of the two samples. Further, the number of degrees of freedom to look up the *p*-value from the resulting *t*-value is the same as that used by the Welch test. This has the satisfying result that when trimming is 0, the Yuen and the Welch test are identical and thus Yuen's test is often referred to as the Yuen–Welch test.

As hoped, the Yuen–Welch test can provide good control of both Type I errors and Type II errors when there are substantial deviations from the normal distribution. It will never be perfect though and, just like a *t*-test, it is better if sample sizes are equal (Wilcox, 2017, p. 170). If the underlying distributions are known to be normal then a Yuen–Welch test with 20% trimming is not as powerful as the Welch test alone (has a higher Type II error rate). The trimming is removing useful data points. But if we were able to know that the underlying data were normal in the first place then there would be no need for any of this!

Though M-estimators offer a wide generalisation of robust measures of location, they do not enjoy the same simple translation back to a *t*-test. Differences in M-estimators can only be converted to *p*-values through a bootstrapping process. Boostrapping is a problem for another day but basically it means using the samples themselves as the best approximations to the underlying distributions. With modern statistical packages, it is no more demanding than a traditional *t*-test.

11.7   Overall Advice

In a usability test, or even some experimental contexts, where what is wanted is a clear sense of the amount of change in location of some measure, be it reduced task time, decreased errors or improved user experience, then there is no need to compare means and apply a traditional *t*-test. Trimmed means, typically with $\gamma = 0.2$, will give good robust measures of location. The Yuen–Welch test will behave, for all practical purposes, like a *t*-test on these trimmed means but is much less likely to give a misleading result when the underlying distributions violate the assumptions of a *t*-test. Though the test itself is more complicated, I think this approach passes my criteria of simplicity and articulation (Chapter 5): it is still very like doing a *t*-test and trimmed means are reasonably straightforward concepts.

Better yet, there is never any need to do a Levene test for homoscedasticity or a test for the normality of distributions, which is just as well because these tests do not match the issues of finding out whether a traditional *t*-test is reliable for the data gathered. And in any case, a Yuen–Welch test will give a good result even if the data are normal. There is a small loss of power if underlying distributions are genuinely normal but that can never be known for sure and if power is a concern then simply collect a few more participants.

Though there is enthusiasm amongst statisticians for M-estimators, I think they do not pass my criteria of simplicity and articulation. First, the choice of ξ function is an important decision but which ξ is best to use depends on the situation. Second, the meaning of the M-estimator, once it has been evaluated, is not straightforward because it is often a complicated function of the data, partially defined on what the data look like for each sample separately. It is also not clear that tests based on M-estimators bring particular advantages over the Yuen–Welch test (Dixon, 1980). I therefore side with Dixon (1980) that in terms of explanation and clarity, the Yuen–Welch test is probably the best all round choice.

References

Algina, James, T. C. Oshima and Wen-Ying Lin (1994). 'Type I error rates for Welch's test and James's second-order test under nonnormality and inequality of variance when there are two groups'. *Journal of Educational Statistics* 19.3, pp. 275–291.

Box, George E. P. (1954). 'Some theorems on quadratic forms applied in the study of analysis of variance problems, I. Effect of inequality of variance in the one-way classification'. *The Annals of Mathematical Statistics* 25.2, pp. 290–302.

Brown, Morton B. and Alan B. Forsythe (1974). 'Robust tests for the equality of variances'. *Journal of the American Statistical Association* 69.346, pp. 364–367.

Dixon, W. J. (1980). 'Efficient analysis of experimental observations'. *Annual Review of Pharmacology and Toxicology* 20.1, pp. 441–462.

Emerson, John D. and Michal A. Stoto (1983). 'Transforming data'. In: *Understanding Robust and Exploratory Data Analysis*. Ed. by David Hoaglin, Frederick Mosteller and John Tukey. John Wiley, pp. 97–128.

Erceg-Hurn, David M. and Vikki M. Mirosevich (2008). 'Modern robust statistical methods: An easy way to maximize the accuracy and power of your research'. *American Psychologist* 63.7, pp. 591–601.

Goodall, Coling (1983). 'M-estimators of location: An outline of the theory'. In: *Understanding Robust and Exploratory Data Analysis*. Ed. by David Hoaglin, Frederick Mosteller and John Tukey. John Wiley, pp. 339–403.

Grissom, Robert J. (2000). 'Heterogeneity of variance in clinical data'. *Journal of Consulting and Clinical Psychology* 68.1, p. 155.

Hill, MaryAnn and W. J. Dixon (1982). 'Robustness in real life: A study of clinical laboratory data'. *Biometrics* 38.2, pp. 377–396.

Pratt, John W. (1964). 'Robustness of some procedures for the two-sample location problem'. *Journal of the American Statistical Association* 59.307, pp. 665–680.

Ramsey, Philip H. (1980). 'Exact type 1 error rates for robustness of student's t test with unequal variances'. *Journal of Educational Statistics* 5.4, pp. 337–349.

Rosenberger, James L. and Miriam Gasko (1983). 'Comparing location estimators: Trimmed means, medians, and trimean'. In: *Understanding Robust and Exploratory Data Analysis*. Ed. by David Hoaglin, Frederick Mosteller and John Tukey. John Wiley, pp. 297–336.

Sawilowsky, Shlomo S. and R. Clifford Blair (1992). 'A more realistic look at the robustness and Type II error properties of the t test to departures from population normality'. *Psychological Bulletin* 111.2, pp. 352–360.

Welch, Bernard L. (1947). 'The generalization of student's' problem when several different population variances are involved'. *Biometrika* 34.1/2, pp. 28–35.

Wilcox, Rand R. (1998). 'How many discoveries have been lost by ignoring modern statistical methods?'. *American Psychologist* 53.3, p. 300.

(2017). *Introduction to Robust Estimation and Hypothesis Testing*. 4th edn. Academic Press.

Yuen, Karen K. (1974). 'The two-sample trimmed t for unequal population variances'. *Biometrika* 61.1, pp. 165–170.

CHAPTER 12

The ANOVA Family and Friends

Questions I am asked:

▷ What exactly does an ANOVA do?
▷ How do I interpret a two-way ANOVA?
▷ I have found a significant four-way interaction effect. Isn't that good?
▷ Isn't ANOVA robust so I don't need to worry about assumptions of normality?
▷ I don't think I can use an ANOVA, but what can I use?

Analysis of Variance, or ANOVA, is the cornerstone of much traditional statistical analysis. Its prevalence arises because of its flexibility to deal with multiple conditions of an independent variable and also multiple independent variables. ANOVAs can deal with within-participants or between-participants designs and any mix of such designs across independent variables. Moreover, the general approach of ANOVA can be extended to analyse even more complex designs including adding covariates (Analysis of Covariance, ANCOVA), using multiple dependent variables (Multivariate ANOVA, MANOVA) or both (MANCOVA). Though firmly rooted in parametric statistics, early analysis suggested that ANOVA was robust to deviations from assumptions, in particular deviations from normality and deviations from homogeneity of variance. The result was that ANOVA has become the workhorse of traditional statistical analysis.

But as we saw with *t*-tests in Chapter 11, violations of assumptions are common in real data and the impact on ANOVA are at least as bad as for a *t*-test because in its simplest form ANOVA and *t*-tests are equivalent. The purpose of this chapter is to look more closely at what ANOVAs actually do and just how robust they actually are. By thinking about what is being tested in ANOVA, that is, the effects that are expected, it is possible to suggest robust alternatives to ANOVA that have many of the advantages of the traditional approaches but greatly reduce their unreliablility when data are not well-behaved and normal. However, when using an alternative to ANOVA it is really important to recognise that the alternative may not allow you to interpret your findings in the way that you would with an ANOVA.

12.1   What ANOVA Does

Like *t*-test, ANOVA is a term that actually covers a family of tests that all work on the same basic principles. The effects that ANOVAs look for are the amount variation of a dependent variable that is due to changes in one or more independent variables. This is why they are analyses of variance. However, because ANOVAs are based on the assumptions of normal data, the calculations in ANOVA are strongly related to means and therefore looking at the amount of variation is equivalent to looking at changes in location. The two distinct types of effect are one and the same in the context of ANOVA.

One implication of ANOVAs actually being a family of tests is that there are different assumptions underlying the tests, the most notable being the different assumptions needed for within-participant and between-participant designs. These different assumptions can be violated in different ways. Thus, when talking about the robustness of ANOVA, we need to be clear about which type of ANOVA we are referring to.

The simplest type of ANOVA is used for analysing a between-participants design with one independent variable. So for instance, Imran Nordin ran a study looking at the effect of lighting levels on people's immersion in digital games (Nordin et al., 2014). The expectation was, fitting with accounts of how people like to play games, that the lower the level of lighting, the more immersed players would feel. Imran's experiment therefore looked at three levels of lighting: dim, normal and bright. As suspected, increasing levels of lighting led to decreasing average immersion scores, but obviously this needed to be supported by statistical testing.

When looking to see the effects of lighting on mean immersion, it is not appropriate to run three *t*-tests on each pair of conditions because this would automatically lead to over-testing (Chapter 13) and the risk of seeing a chance significant result (a Type I error) above the expected chance of $\alpha = 0.05$. This is not to mention that multiple testing also reduces the severity of a severe test. To avoid these problems, an ANOVA is used because it is an omnibus test, that is, it tests all three conditions in one go.

A significant result from an ANOVA is typically interpreted as meaning that there is some difference between the means of the three conditions. What the ANOVA cannot do is say exactly what that difference is. For instance, a significant result in Imran's experiment could mean that immersion in the dim condition was indistinguishable from the normal condition but different from the bright condition or equally well that all three conditions were meaningfully different from each other. The cost of an omnibus test is that you find out that there is a difference but not what the difference is.

In all types of ANOVA, what the test does is compare variations in means across experimental conditions with the natural variation of individuals within each experimental condition. So with the lighting study, the natural variation of individuals is captured by the three different standard deviations of immersion scores for each of the three conditions of the experiment. These are pooled to give an overall measure of standard deviation of individuals. In fact, the variance, the square of the standard deviation, is used hence the term Analysis of Variance. The variance of the three mean immersion scores for each group[1] is also calculated and compared with the variance of individuals. This is illustrated in Table 12.1. When the variation due to means is substantially more than the natural variation of individuals, indicated by a large *F* statistic, then the ANOVA gives a significant *p*-value. And the reasoning then goes that the difference in means is due to the experimental manipulation and not natural variation, and hence we would have evidence that lighting does influence the immersive experience of players.

For a repeated-measures one-way ANOVA, the process is similar but instead of looking at differences in means for each group, the differences between an individual's scores are used. The null hypothesis would be that the mean of the differences are all zero. In this case, the ANOVA compares variation in mean *differences* to natural variation in the *differences*. When the data meet the assumptions of ANOVA, we can move freely between means and variances so, in fact, do not directly consider the variances of differences. Instead, the repeated-measures ANOVA calculation use these three sources of variance (following the notation of Table 12.1):

- the pooled variance of the scores in a particular condition, v_p
- the variance of the means of each condition, v_M
- and also the variance of the mean scores of each participant, v_S (S for subject!)

---

[1] Yes, this is a variance of just three items so there's not a lot of variation to see!

Table 12.1. *The different sources of variation in a one-way between-participants ANOVA where X_i represents the data point gathered from a single participant.*

	Condition 1	Condition 2	Condition 3
Scores	X_a X_d $\vdots$	X_b X_e $\vdots$	X_c X_f $\vdots$
Mean of each condition	M_1	M_2	M_3
Variance of Means		v_M	
Variance of each condition (standard deviations squared)	v_1	v_2	v_3
Pooled Variance		v_p	
***F*-value**		v_M/v_p	

This last source of variance reflects that there are individual differences between people that are unrelated to the conditions for the experiment and recognising this helps to improve our estimate of whether v_M is a lot more than might be expected by natural variation alone.

Even though the calculation is quite different, the interpretation of a repeated-measures ANOVA is similar to that of a between-participants ANOVA. A large *F*-value means that there is more variation in scores between conditions than would be expected by natural variation of individuals and hence that the experimental manipulation is affecting people's scores.

With a two-way ANOVA, there are two independent variables and so more sources of variation. For instance, having found that lighting has an effect in the previous example study, we might devise a further study not only where lighting is either low or bright but also where participants wear headphones or not as a further level of dissociation from the real world. Each variable alone might contribute variation so the ANOVA gives a main effect for each variable and corresponding *F* and *p*-values. That is, headphones alone might influence immersion and this would be indicated by the variance of the two mean immersion scores based solely on whether participants wore headphones or not (averaging across both lighting conditions). Similarly, there would be a main effect for lighting, irrespective of whether participants wore headphones. The further analysis that a two-way ANOVA can give is whether there was an interaction between the two variables so that in some combination the variables produced a further effect not attributable to each item alone. This is the interaction effect and the ANOVA produces a further F and p-value for this.

Interaction effects are quite tricky to think about and often not interpreted carefully enough (Rosnow and Rosenthal, 1991). It is not simply that some combination of the variables produces an effect but that there is effectively some sort of crossover effect having removed any differences due to main effects. For example, suppose there were no main effects for either headphones or lighting in our extended study but there was an interaction. The interaction would mean that in low light *with* headphones, immersion was about the same as in bright light *without* headphones but different from the other two conditions of low light without headphones and bright light with headphones. Furthermore, these two conditions should similar to each other. This is not the typical interpretation of an interaction effect.

The best way to get used to what two-way ANOVAs do is to look at interaction plots of the means produced in each experimental condition. The interaction plots when there is only one effect and all other effects are zero are illustrated in Figure 12.1. Notice in particular that a two-way

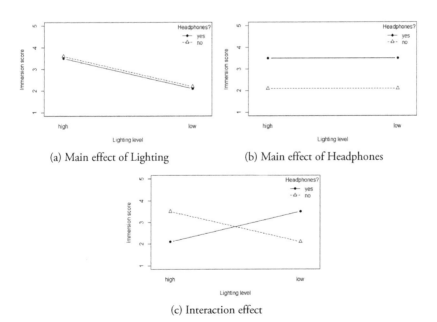

(a) Main effect of Lighting   (b) Main effect of Headphones

(c) Interaction effect

Figure 12.1   Hypothetical interaction plots for one effect in a two-way ANOVA when both other effects are zero.

interaction is only found if there is an actual crossover as shown in Figure 12.1c. However, when there are also main effects, the crossover can be harder to see and even harder to interpret, and harder still when there are differences in main effects that are not significant. This is one of those situations where you have to make sure your experimental design is simple enough for you to be able to interpret the results (Chapter 5).

The ANOVA family can keep on going to include arbitrary numbers of independent variables. But this has two drawbacks. First, the more independent variables there are, the more F and p-values produced and this means that over-testing reappears. It may still not be as bad as pairwise testing of every experimental condition but it is still over-testing (see Chapter 13 for a full discussion). More importantly, the larger the number of variables, the less severe a test. What exactly is being tested if you think you need four distinct experimental manipulations to test it?

Even if you are clear about the severity of the test or if you are just exploring your data, explaining a higher order interaction effect is very challenging. It may be that you suspect there is some interaction between all the variables but higher order interaction effects only indicate differences having already accounted for lower order interactions and main effects. Thus, a three-way interaction effect actually reflects a difference in the two-way interactions of two variables between the levels of the third variable. For example, suppose we had a third variable in the lighting condition, say the level of horror content in the game. A three-way interaction in this context would mean that there is a difference in the *interaction* between Headphones and Lighting when the levels of Horror differed. Moreover, this difference would be independent of any main effects of each variable, or even an overall interaction effect of Headphones and Lighting.

Similarly, a four-way interaction effect indicates a difference in the three-way interactions for different levels of a fourth variable so therefore represents a change in the degree of the change in degree of crossover of two of the variables. I think. I'm not wholly sure that is correct as I write it. And even if you are, what would it mean? It fails my criterion of articulation (Chapter 5). And that's just in its simplest form. As soon as there are main effects and lower order interaction effects, these interpretations become even more complicated.

For these reasons, I do not advocate going beyond two-way ANOVAs and I do not consider them further here. One-way and two-way ANOVA designs are sufficient for so many practical purposes. Even if you are convinced that a three-way interaction is important, there are real challenges in interpreting it in any meaningful way. You might be better off trying to design a better study.

12.2   Is ANOVA Robust?

The general claim is that ANOVA is robust to deviations from its underlying assumption. The sources for this claim go back some way and typically start with Box (1954) but there is a big gap between the general claims of robustness and what was actually found. The assumptions for the simplest ANOVA, a between participants ANOVA on one independent variable, are similar to those for a *t*-test (Chapter 11) but they are worth restating here:

1.   Independence of observations
2.   Normality
3.   Homoscedasticity

The first assumption is a standard of all tests. It means that results of one participant are independent of the results of any other participant and is usually met by experimental design. It is not a particular concern for the robustness of the test, so I won't consider it further except to include it for completeness.

What Box (1954) actually considered was data that were normally distributed but violated the assumption of homoscedasticity, that is, the variance (standard deviation) of the data varied substantially between groups in the ANOVA. What Box found was that when the ANOVA was balanced, which is to say that groups were all the same size, ANOVA maintained a Type I error rate when homoscedasticity was violated. This sounds very promising: Type I error rate is not everything in modern statistical analysis but it is a useful measure of the robustness of a test (Chapter 9). However, Box was working with quite modest differences in variance, typically a ratio of 3:1 in the most extreme cases. But this is a very modest range of variances, corresponding to a ratio of standard deviations of less than 2:1. Real datasets exhibit much greater differences (Wilcox, 1987) and when the ratio of standard deviations (rather than variances) is 3:1 or 4:1, then ANOVA loses control over the Type I error rate even with quite respectable sample sizes of more than 10 participants per group and even when the groups are balanced for size (Brown and Forsythe, 1974; Wilcox, 1987; Coombs, Algina, and Oltman, 1996). Box's (1954) original analysis was not enough to address practical issues of robustness with datasets that arise in many realistic situations.

When it comes to the violation of the assumption of normality, it seems at first glance that ANOVA is actually robust. When there is the assumption that all the distributions are equal except for location, then ANOVA shows both good control of Type I errors and good power. An early review is given in Tan (1982) and more recent studies show similar results

(Khan and Rayner, 2003; Schmider et al., 2010). However, in practice, the distributions are not always equal between conditions, in particular, skew and variance tend to change as the mean also changes. In these situations where the two assumptions are violated, ANOVA simply detects differences in distributions, which is not systematically related to any difference in means (Zimmerman, 1998). Arguably, in such situations, ANOVA is correct in detecting the difference in distributions but the usual interpretation as a difference in means would be unsound. Note that the same problems of interpretation are seen with non-parametric tests (Chapter 10) and, in particular, when considering non-parametric alternatives to ANOVA as discussed below.

These are the problems just for between-participants ANOVA. For repeated-measures ANOVA, the problems are more complex. Because each person is measured multiple times, this forms a multivariate sample for which the assumptions are:

1. Multivariate normality
2. Sphericity
3. Independence of observations

Multivariate normality is not just the assumption that the data are normal within each experimental condition but that as a whole, the data come from a multivariate-normal distribution. This is a considerably stronger assumption than normality in between-participants designs as it requires not only the normality of the variable within each condition but also the normality of every (linear) combination of measurements.

Sphericity corresponds to the homoscedaticity assumption of between-participant ANOVA; however, it is referring to the covariance structure (roughly a combination of the variances and correlations between the measures made on the participants). Without getting too far into technicalities, it is a much stronger assumption than homoscedasticity and correspondingly much more easy to violate. A good review of the issues in repeated-measures ANOVA, and in particular the sphericity assumption, is given in Keselman, Algina, and Kowalchuk (2001).

Early studies suggested that repeated measures ANOVA is robust to deviations from multivariate normality (Keseleman, Lix and Keselman, 1996) However, later studies looked at stronger variations from normality and showed a considerable lack of robustness (Oberfeld and Franke, 2013). Even when data is suitably normal, repeated measures ANOVA is nonetheless very sensitive to violations of sphericity, moreover these violations are fairly common (Keselman, Lix, and Keselman, 1996). Fortunately, unlike violations of homoscedasticity (Chapter 11), there is a test, Mauchly's test, that is able to detect violations of sphericity, which is meaningful in the context of an ANOVA. In these cases, the degrees of freedom for converting *F*-values to *p*-values can be adjusted much as the Welch test adjusts a *t*-test for lack of homoscedasticity (Chapter 11). The two most commonly used corrections are called the Huynh–Feldt and the Greenhouse–Geisser corrections and they work well provided the data are normal (Cardinal and Aitken, 2013).

One further complication comes from a two-way mixed design (also called a split-plot design) where there is one repeated measure and one between-participants variable. For this situation, there is a further assumption, that sphericity of the within variable is homogeneous across the between variable, and a mixed-measures ANOVA is sensitive to violation of that assumption as well (Keselman, Algina, and Kowalchuk, 2001).

With all this evidence accumulating, it seems that the claim for the robustness of ANOVA is really not tenable any more. Simple, commonly seen features of real data, such as differences in variance or lack of sphericity, severely undermine the reliability of traditional ANOVA. Add into consideration that changes in these factors often come with changes in the shape of the underlying distributions, and the two violations together make the sound interpretation of ANOVA hazardous, to say the least. The analyses of robustness reported here have not even considered issues due to outliers. Fortunately, as with much other traditional statistics, new methods are emerging and they present straightforward alternatives to ANOVA, but they come at a cost of interpreting the results.

12.3   Robust Alternatives to ANOVA

When thinking of alternatives to using an ANOVA, it is important to remember that ANOVA is interpreted as a test of differences in means between groups and so as such is looking for differences in location between the conditions (Chapter 4). But this is really because when the assumptions of ANOVA are met, it is safe to move from effects based on variance to effects based on changes in location. This is not true of tests that do not assume well-behaved normality.

12.3.1   Non-Parametric Alternatives

For one-way designs with a single independent variable, when it is known that ANOVA is unsuitable, it is common practice to use instead the traditional rank-based non-parametric tests: Kruskal–Wallis for between designs and Friedman for repeated-measures. For repeated-measures, the Friedman is a good choice. It shows good control of Type I errors and reasonable power (Zimmerman and Zumbo, 1993) but, and this is important, it detects differences in the dominance between groups, not differences in location. This means that when there are radically different distributions between the conditions, the Friedman can give significance reliably but not because of the difference in means or medians (St. Laurent and Turk, 2013). There are also other rank-based methods that improve on the power of the Friedman test somewhat but are also only testing the hypothesis of differences in distributions. These are discussed more fully in Wilcox (2017, p. 447).

By contrast, the Kruskal–Wallis is not so reliable though it may be sufficient in many practical situations as discussed in Chapter 10. It suffers from some of the problems of ANOVA, which is that when there is not homogeneity of variance the Kruskal–Wallis test becomes unreliable, even when the groups are balanced in terms of size (Lix, Keselman, and Keselman, 1996). However, like the Friedman, if the Kruskal–Wallis is used to test for differences in distribution, it is very reliable, and more so than ANOVA when sample sizes are reasonable (15 or more per group) (Khan and Rayner, 2003).

For two-way ANOVA designs, it might be tempting to simply substitute ranks of scores for actual scores and perform the ANOVA on the ranks. This, after all, is more or less what the Mann–Whitney and Wilcoxon tests do only with regards to t-tests (Conover and Iman, 1981). However, this naive approach has been shown to have problems. The testing of the interaction effect can have very high Type I error rates when there are main effects (Beasley, 2002) and poor power (Sawilowsky, Blair, and Higgins, 1989), in some cases worse than the ANOVA it is meant to replace.

One approach that is proposed to improve on this but still work with ranks is called Aligned Rank Transforms. These have come to prominence in HCI (and more widely) due to an exposition and supporting software produced by Wobbrock et al. (2011). In these tests, it is recognised that all effects except the specific one of interest are potentially nuisance parameters when it comes to ranking. So to avoid this, before ranking, the data are aligned by subtracting differences due to the other effects. So the Wobbrock et al. (2011) approach, in our lighting and headphones example of Section 12.1, would be as follows. To test for the interaction effect, take each data point and subtract the corresponding main effect means. For example, for someone who was measured in low light without headphones, subtract the mean of *all* the low lighting scores and the mean of *all* the without-headphones scores. Thus, the scores in that group have been aligned by subtracting the main effects. Similarly, the scores in the other groups are aligned by subtracting their corresponding means for main effects. All of these aligned scores are then ranked and an ANOVA performed, but only the interaction effect is considered as that is the only one that has not been aligned.

There are different mechanisms for aligned rank tests. Ranks can be aligned not only on means but also on medians. Test statistics can be compared to different thresholds, not just the standard F values, either by considering different distributions or adjusting the degrees of freedom (O'Gorman, 2001; Toothaker and Newman, 1994). However, regardless of the exact details, though slightly more complicated, the transformation is not so hard to understand and offers some of the simplicity of rank-based statistics. Moreover, it shows improvement in robustness over correspond- ing traditional parametric approaches (O'Gorman, 2001; Beasley, 2002). However, there are situations where the robustness does not come through and aligned rank tests can show lack of power or lack of control of Type I error (Toothaker and Newman, 1994).

There is also the issue of interpretation. For instance, once alignment has been done, the tests are comparing ranks and therefore comparable to traditional Mann–Whitney tests, Friedman tests and their like. However, these tests are better considered as tests of dominance rather than tests of location (Beasley, 2002). They are therefore sensitive to differences in the data other than difference in location (also discussed in Chapter 10), in particular, differences unrelated to the means used to align them. Thus, while conceptually the tests seem close to rank-based tests, their interpretation feels rather ambiguous to me, being somewhere between a location test and a test of dominance.

There are also other rank-based tests in the literature that are potential alternatives to two-way ANOVA, as described by Wilcox (2017). And con- ceivably, there are many more possible by combining different alignment procedures, say with robust measures of location and with different ways of comparing ranked data. I am unaware of a systematic comparison of these tests to either traditional ANOVA tests or existing alternatives to ANOVA so I cannot recommend what would be most reliable to work with.

12.3.2 Changes of Location

If your focus really is on changes of location then a good alternative is a variation of ANOVA that works with trimmed means. The concerns and practicalities are identical to those for robust alternatives to *t*-tests and I refer you Chapter 11 for a detailed discussion of using trimmed means. There is an alternative for between-participants ANOVA based on the Welch variation on the *F*-test but also adapted for analysing trimmed means. This is the ANOVA version of the Yuen–Welch test. It has been shown to provide good robustness in the face of both non-normality and heteroscedasticity (Lix and Keselman, 1998; Cribbie et al., 2012). Similarly, for repeated-measures designs, an approach based on the standard Huynh–Feldt correction for violation of sphericity and again working with trimmed means shows good robustness (Wilcox, 1993).

For two-way designs, whether between, within or mixed, the story is quite similar with regards to examining differences in location. There are variants of ANOVA building on the Welch test that have been developed for these more complex designs that are robust to violations of sphericity and homoscedasticity (Johansen, 1980). When these are further adapted to use trimmed means as robust measures of location, the resulting tests are robust to violations of homoscedasticity, sphericity, homogeneity of sphericity and non-normality (H. Keselman, Algina, Wilcox, et al., 2000; Kowalchuk, H. Keselman, and Algina, 2003; Wilcox, 2017). These present good alternatives to ANOVA when the focus of concern really is a change in location. Moreover, they seem to show better robustness compared to aligned rank tests (Kowalchuk, H. Keselman and Algina, 2003).

More practically, these alternative tests for differences in location are included in the WRS2 package for R and both the tests and examples of the R code are given in Wilcox (2017).

12.3.3   Do Something Else

Even though there seem to be some good alternatives to ANOVA in many situations, it has to be said that no alternative will be robust in all situations. Rather than losing yourself in the debates about which alternative is best in which situation, it might sometimes be easier simply to avoid the issue entirely by doing something else.

For example, consider a pre- and post-test style of experiment where it is hoped that students feel more confident going into an interview having used a new interactive learning resource compared to having used a traditional resource. The issue is not whether the students felt more confident after the experiment from before the experiment, which would be the main effect of pre- and post-, but rather whether the differences in confidence are bigger for those who used the new resource. Thus, it would be enough to simply test the differences in confidence rather than conduct a mixed-measures ANOVA. Confidence before the experiment would be checked just to make sure that we did not happen, by chance, to have a substantially more or less confident group of people in one of the experimental conditions, as that might skew our ability to detect effects.

Alternatively, if you have multiple conditions, do you really need them all? If your research is at an early stage, it might be better simply to consider what you expect to be the more extreme conditions. In Imran's lighting experiment, did we really need low, medium and bright lighting conditions, or could we just have got away with low and bright? Similarly, unless you really are interested in interactions between variables, it might be better to test the effect of two variables separately: headphones in one experiment and lighting in another before trying to see the effect of both at once.

By simplifying the design of the study, you are asking more focused questions and this tends to lead to a more severe test. If, however, such complexity is necessary, another way to be focused is to use what is called a contrast. Here, over a set of conditions, our analysis of the situation suggests a specific pattern of differences between the conditions. In the lighting experiment, it was expected that as lighting went up, immersion would go down. This is a very specific pattern of change over the three conditions. We did not reduce the number of conditions because it was not clear to us where the lighting would begin to have an effect on players' experiences. Contrasts fit very well with severe testing as they test a highly specific hypothesis. The cost is that if there is a difference, but it does not fit that hypothesis, the contrast will not see it at all. I make the analogy between a lantern and a torch. While both may use the same strength light bulb, the lantern is an ANOVA, spreading the light evenly around you so it illuminates interesting features in all directions. The torch is a contrast, pointing all the light in only one direction so you can see much further if there are interesting features in that direction but you'll not see the interesting features that might be much closer to you if they are not in the beam.

12.4 Summary

The appeal of ANOVAs is clear. They allow for more complicated analyses including the interaction between variables, while at the same time controlling for over-testing. The problem is that they are firmly rooted in the parametric style of statistics and, moreover, their purported robustness to deviations from assumptions is not really supportable in the face of realistic data produced in experiments. Fortunately, there are alternatives. For one-way analyses, rank-based methods offer familiar and acceptable tests but the cost is that they are not necessarily testing changes in location, merely changes in the distributions of the data. Where location is the important consideration, there are robust alternatives based on trimmed means that offer all the flexibility of traditional ANOVA but have much better behaviour across a wide range of realistic experimental situations. But occasionally, it is worth stepping back from a fixation on ANOVA and wondering if it might be possible to do something else entirely. After all, there is no imperative to make our experiments match a particular style of statistical testing.

References

Beasley, T. Mark (2002). 'Multivariate aligned rank test for interactions in multiple group repeated measures designs'. *Multivariate Behavioral Research* 37.2, pp. 197–226.

Box, George E. P. et al. (1954). 'Some theorems on quadratic forms applied in the study of analysis of variance problems, I. Effect of inequality of variance in the one-way classification'. *The Annals of Mathematical Statistics* 25.2, pp. 290–302.

Brown, Morton B. and Alan B. Forsythe (1974). 'The small sample behavior of some statistics which test the equality of several means'. *Technometrics* 16.1, pp. 129–132.

Cardinal, Rudolf N. and Michael R. F. Aitken (2013). *ANOVA for the Behavioral Sciences Researcher*. Psychology Press.

Conover, William J. and Ronald L. Iman (1981). 'Rank transformations as a bridge between parametric and nonparametric statistics'. *The American Statistician* 35.3, pp. 124–129.

Coombs, William T., James Algina and Debra Olson Oltman (1996). 'Univariate and multivariate omnibus hypothesis tests selected to control type I error rates when population variances are not necessarily equal'. *Review of Educational Research* 66.2, pp. 137–179.

Cribbie, Robert A., Lisa Fiksenbaum, H. J. Keselman and Rand R. Wilcox (2012). 'Effect of non-normality on test statistics for one-way independent groups designs'. *British Journal of Mathematical and Statistical Psychology* 65.1, pp. 56–73.

Johansen, Søren (1980). 'The Welch-James approximation to the distribution of the residual sum of squares in a weighted linear regression'. *Biometrika* 67.1, pp. 85–92.

Keselman, H. J., James Algina and Rhonda K. Kowalchuk (2001). 'The analysis of repeated measures designs: a review'. *British Journal of Mathematical and Statistical Psychology* 54.1, pp. 1–20.

Keselman, H. J., James Algina, Rand R. Wilcox and Rhonda K. Kowa (2000). 'Testing repeated measures hypotheses when covariance matrices are heterogeneous: Revisiting the robustness of the Welch-James test again'. *Educational and Psychological Measurement* 60.6, pp. 925–938.

Keselman, Joanne C., Lisa M. Lix and H. J. Keselman (1996). 'The analysis of repeated measurements: A quantitative research synthesis'. *British Journal of Mathematical and Statistical Psychology* 49.2, pp. 275–298.

Khan, Azmeri and Glen D. Rayner (2003). 'Robustness to non-normality of common tests for the many-sample location problem'. *Journal of Applied Mathematics & Decision Sciences* 7.4, pp. 187–206.

Kowalchuk, Rhonda K., H. J. Keselman and James Algina (2003). 'Repeated measures interaction test with aligned ranks'. *Multivariate Behavioral Research* 38.4, pp. 433–461.

Lix, Lisa M. and H. J. Keselman (1998). 'To trim or not to trim: Tests of location equality under heteroscedasticityand nonnormality'. *Educational and Psychological Measurement* 58.3, pp. 409–429.

Lix, Lisa M., Joanne C. Keselman and H. J. Keselman (1996). 'Consequences of assumption violations revisited: A quantitative review of alternatives to the one-way analysis of variance F test'. *Review of Educational Research* 66.4, pp. 579–619.

Nordin, A. Imran, P. Cairns, M. Hudson, A. Alonso and E. H. Calvillo Gamez (2014). 'The effect of surroundings on gaming experience'. *Foundations of Digital Games*. Available at: www.fdg2014.org/papers/fdg2014_wip_14.pdf [Accessed October 1, 2018].

Oberfeld, Daniel and Thomas Franke (2013). 'Evaluating the robustness of repeated measures analyses: The case of small sample sizes and nonnormal data'. *Behavior Research Methods* 45.3, pp. 792–812.

O'Gorman, Thomas W. (2001). 'A comparison of the F-test, Friedman's test, and several aligned rank tests for the analysis of randomized complete blocks'. *Journal of Agricultural, Biological, and Environmental Statistics* 6.3, pp. 367–378.

Rosnow, Ralph L. and Robert Rosenthal (1991). 'If you're looking at the cell means, you're not looking at only the interaction (unless all main effects are zero)'. *Psychological Bulletin* 110.3, pp. 574–576.

Sawilowsky, Shlomo S., R. Clifford Blair and James J. Higgins (1989). 'An investigation of the type I error and power properties of the rank transform procedure in factorial ANOVA'. *Journal of Educational Statistics* 14.3, pp. 255–267.

Schmider, Emanuel, Matthias Ziegler, Erik Danay, Luzi Beyer and Markus Bühner (2010). 'Is it really robust?'. *Methodology* 6, pp. 147–151.

St. Laurent, Roy and Philip Turk (2013). 'The effects of misconceptions on the properties of Friedman's test'. *Communications in Statistics-Simulation and Computation* 42.7, pp. 1596–1615.

Tan, W. Y. (1982). 'Sampling distributions and robustness of t, F and variance-ratio in two samples and ANOVA models with respect to departure from normality'. *Comm. Statist.-Theor. Meth.* 11, pp. 2485–2511.

Toothaker, Larry E. and De Newman (1994). 'Nonparametric competitors to the two-way ANOVA'. *Journal of Educational Statistics* 19.3, pp. 237–273.

Wilcox, Rand R. (1987). 'New designs in analysis of variance'. *Annual Review of Psychology* 38.1, pp. 29–60.

(1993). 'Analysing repeated measures or randomized block designs using trimmed means'. *British Journal of Mathematical and Statistical Psychology* 46.1, pp. 63–76.

(2017). *Introduction to Robust Estimation and Hypothesis Testing*. 4th edn. Academic Press.

Wobbrock, Jacob O., Leah Findlater, Darren Gergle and James J. Higgins (2011). 'The aligned rank transform for nonparametric factorial analyses using only anova procedures'. *Proc. of CHI 2011*. ACM. ACM Press, pp. 143–146.

Zimmerman, Donald W. (1998). 'Invalidation of parametric and nonparametric statistical tests by concurrent violation of two assumptions'. *The Journal of Experimental Education* 67.1, pp. 55–68.

Zimmerman, Donald W. and Bruno D. Zumbo (1993). 'Relative power of the Wilcoxon test, the Friedman test, and repeated-measures ANOVA on ranks'. *The Journal of Experimental Education* 62.1, pp. 75–86.

CHAPTER 13

Exploring, Over-Testing and Fishing

Questions I am asked:

▷ I've got all these variables that might be important so what is the best experiment to do?

▷ My result isn't significant: what should I do?

▷ Why does it matter how many tests I do?

▷ I've got all these significant results. Which ones should I report?

On the whole, gathering data is hard work. So whenever you run a study to answer a particular question and gather data that should answer it, you also, first, gather data that tell you about the context of the study, a common example of which is gathering participant details. Second, it is also tempting to gather other data that may or may not be entirely relevant to the question but may nonetheless be interesting; for instance, in a study looking at the effect of a new menu structure on people's success in completing a task, you may also gather timings of how long people take between clicks of the mouse, where their eyes were looking during the task or even interview them about their experiences of menus generally. All of these data offer the chance to explore more about people, perhaps find out more about your original research question and perhaps to set off new lines of inquiry. The more data, the better!

However, there is a fine line that needs to be walked here. Exploring data is one thing. Over-testing is another. And fishing for results is yet another thing entirely. In this chapter, I discuss the differences between these approaches and how you might find yourself slipping from exploration into bad practices, and some rules of thumb that might help you to avoid this. I also make the distinction between exploring data arising from a severe test, that is, an experiment with a specific hypothesis in mind, and a study that sets out to be purely exploratory even though it looks like an experiment.

13.1   Exploring After a Severe Test

Conducting an experiment as a good severe test is a worthy goal in advancing knowledge, as discussed in Chapter 1. Within the argument of severe testing, there is a single statistical test that is used to evaluate whether or not the experimental hypothesis has been supported and that primary test is the goal at the heart of the planning, conduct and analysis of the experiment. However, it does not do justice to your efforts to simply do the primary test and stop. All experiments are necessarily limited and narrow in scope and so a critical reader, of which you would *want* to have many, might look at the experiment and raise an objection to your interpretation.

For example, in Alena Denisova's work on the placebo effect in digital games (Denisova and Cairns, 2015), in one experimen she deliberately misinformed participants about the quality of the AI in the game. An objection might be that some of her participants were knowledgeable about AI and so not so easily fooled by her misinformation. And it might be that, by chance, there were more AI-knowledgeable people in one condition than another. This would mean that what looked like people being influenced by the information they were given was in fact due to the fact that some people had more information in their heads before they even started the experiment. The typical argument is that randomisation removes any such systematic bias, but when there are potentially dozens of such confounding variables, there can be no guarantee that randomisation has removed any particular confound. Royall (1976) criticised experimenters for failing to look for problems in their samples: "First you randomise then you closurise" (you need to say this out loud).

As ever, one response to this is to run another study to test this specific hypothesis. But as this is a challenge to the specific details of this experiment, a better response is to see whether indeed such a chance allocation of participants did actually happen: open-your-ise! This is what Alena did and first found an even distribution of knowledgeable and unknowledgeable participants between the experimental conditions. Thus, the systematic difference between conditions could not be caused by such a confound. At the same time, she also found that those who were knowledgeable were less influenced by the information she fed them. This difference was statistically significant. So the criticism is valid but did not challenge the interpretation of the primary experimental results.

Effectively, what is happening here is that you, the researcher, are entering into a putative dialogue with your readers, addressing their concerns before they have a chance to voice them. Some of these concerns may well have been raised for you by supervisors, colleagues and reviewers before your work gets published. In your discussion of the study, you lay out these challenges for the reader so they do not have to go over this already established ground again. However, it should also be recognised that such secondary analysis is not a severe test, because the study is not set up around this analysis. Even though Alena's secondary analysis was significant and with an appreciable effect, this is not strong evidence that knowledge of AI influences how people are affected by the information they receive. That would have required her to devise a different study. All that this sort of analysis can say is that some confound was not obviously undermining the severity of the test of the experimental hypothesis.

Exploratory analysis as part of a severe test is, therefore, really a way of building up confidence in the severity of the test. It can never really provide evidence for a different idea, but it might point to ideas that need their own severe testing. And even though the secondary analysis is useful, the best approach is always to run more experiments so as to provide more evidence, which is of course what Alena also did.

13.2   Exploratory Studies

In general, an exploration of an idea can be anything, be it an anecdotal observation, a carefully reasoned philosophical argument, a rich ethnography or even a well-controlled experiment. What I mean here by exploratory *study* is one that looks very much like an experiment where there are named independent variables and quantitative dependent variables and the analysis looks very much like a typical statistical analysis.

The difference between this sort of study and a gold-standard severe test is that there may not be any particular hypothesis under test. For instance, there is currently a trend in player-experience research that individual differences, also called personality types, are an important factor in the types of experiences that players of digital games both seek out and experience. This seems reasonable. The games that people choose to play and why they play them are undoubtedly informed by taste. I would never choose to play a first-person horror shooter game (being a delicate flower) and I know there are plenty of serious gamers who would scoff at my delight in playing *Find the invisible cow*.[1] Accordingly, the individual differences between players should influence many aspects of players' experiences both generally and

---

[1] …because it isn't going to find itself.

with particular games. But which aspects of personality are important and for which experience?

This question lends itself to an exploratory study because there are many well-defined, much-used and reliable measures of personality, most notably the Big Five (Costa Jr. and McCrae, 1992), and there are similarly well-established measures of different aspects of user experience. Moreover, because these concepts are well defined, there are plausible connections: increased extroversion in the player may lead to increased social presence with team-mates; players who are more open to experience may become more immersed in fantasy role-playing games. But are these actual connections, and what other connections of this sort might exist? An exhaustive set of experiments to severely examine each plausible hypothesis would be enormous and likely never completed. Instead, a sensible first step is to conduct an exploratory study, for example, have a set of players play three or four different games and measure their personalities and a range of player experiences.

Such a study looks like an experiment but clearly is not designed to test a particular hypothesis. It is open to a range of possible relationships between games, personalities and experiences. And while not a severe test, it could help to start to see what are the plausible links between the three types of variables. Moreover, the design of the study lends itself to statistical analysis. Typical tests that can deal with such a mix of independent and dependent variables, like ANOVA (Chapter 12), might be used to see the differences in player experiences between the games. Personality measures can be added in as covariates (analysed with ANCOVA).

Rather than use more sophisticated tests like ANCOVA, which are not easy to interpret, pseudo-independent variables can be made. This is when something that is not actually in the researcher's control is treated as if it were an independent variable in the statistical analysis. A typical pseudo-independent variable might be the gender or culture of a participant. An alternative type of pseudo-independent variable is to take something like the personality trait extroversion and, rather than treat it as a continuous variable, divide people into High Extroversion or Low Extroversion. This is typically done two ways, with either a median split or a tercile split. In a median split, the median value of Extroversion for the participants would be the threshold for declaring people to be High or Low. However, there could be a group of people near the median who are not really that different from each other in Extroversion but are classified differently. A split around the median (or any other value) is too simplistic. To overcome this, a tercile split divides participants into thirds according to their Extroversion score. The lowest third are declared Low and the highest third High. This helps to reduce the similarity between Low and High Extroverts but it means that the middle third have to be removed from the analysis so substantially reduces the power of any analysis. The new variable is treated like a normal independent variable and could be included in an ANOVA.

Regardless of how variables are included in the analysis, the results are essentially a form of correlation. Personality might look like it is influencing player experience but equally well, there could be some other unmeasured variable that is influencing both. Correlation is not causation (Chapter 14). Furthermore, without any prior expectation on what associations might be found, any finding from the analysis could still only be chance and not constitute strong evidence. An exploratory study is just that. It helps to explore what might be going on. If there are strong systematic effects then they ought to be seen but at the same time, lots of other chance effects might also show up as false positives. Finding something through an exploratory study is really a way to reduce the search for meaningful connections, but it can only ever be the start of the search.

13.3   Over-Testing

Exploration of data using statistics is a form of analysis distinct from the evidential argument form of severe testing. Whether exploring as the follow-up analysis to discuss the results of a severe test or as the broad-brush investigation into some suspected phenomenon, in both cases there is no specific prediction of how data should turn out so all 'significant' results must have the same status as chance variations.

This is most clearly seen in the reasoning of the traditional NHST approach (Chapter 2) with its emphasis on p-values. A p-value is the probability of an event occurring purely by chance, but if you are happy to accept a large set of events as interesting then you would expect purely by chance to get low p-values for some of the interesting events. Or put another way, if you want to get a double six, the more you roll your pair of dice, the more likely you are to get one.

This problem is called over-testing. The more statistical tests that are run, the more likely it is to get a Type I error of seeing statistical significance when there is no effect. There are various ways to address over-testing.

13.3.1　ANOVA Can (Sometimes) Help

Some statistical tests are designed to sidestep over-testing. These are the omnibus tests, like ANOVA and Kruskal–Wallis. An ANOVA, for instance, is intended to compare multiple conditions in 'one shot' rather than do multiple pairwise comparisons that would lead to over-testing. Multiple comparison are only conducted if the ANOVA turns out to be significant. In this way, the omnibus acts as a gatekeeper to further testing and so protects the family-wise α, that is, it makes sure that when multiple comparisons are done and give significance, it is not because of over-testing. However, this only works for certain statistical designs.

ANOVA comes into its own for exploration because it is able to do comparisons across multiple independent or pseudo-independent variables. For example, á two-way ANOVA compares the differences in some dependent variable in relation to two independent variables (Chapter 12). The results are main effects that see differences due to a single independent variable and an interaction where it is only the combination of two independent variables that leads to an effect. Better yet, having more levels (conditions) of each independent variable does not affect the ANOVA results: a two-way ANOVA tests for two main effects and an interaction effect whether each independent variable has 2 levels (a 2 × 2 design) or 5 (5 × 5 design).

ANOVA also scales up for independent variables. The same sort of analysis can be done for a three-, four- or five-way ANOVA and, alongside main effects, such ANOVAs analyse increasingly complex interactions between the indepedent variables. For instance, with a five-way ANOVA, there can be five four-way interactions where four independent variables in combination influence the dependent variable in combination *having already accounted for any three-way, two-way or main effects*. I do not know about you but I have no idea what a four-way interaction means in any practical context.

Interpretation aside, for an *n*-way ANOVA there are $2^n - 1$ *F* values produced, each corresponding to a separate *p*-value and thus over-testing sneaks in by the backdoor. The ANOVA cannot guard against over-testing in this way: it is designed to guard against over-testing between multiple pairs of conditions within a variable, not between multiple variables. A careful researcher might realise this and consider only main and two-way effects arising from, say, a five-way ANOVA but this would still be 15 separate, independent *p*-values, any one of which might be significant by chance. The results of such an analysis would likely give at least one significant comparison with $p > 0.5$, which is better than flipping a coin.

Of course, the researcher might get several significant results but how would they know which ones were the result of meaningful differences and which were just accidents of the analysis?

13.3.2 Planned Comparisons

An alternative is to plan a specific set of tests. This goes back to making specific predictions across the set of experimental conditions and so severely testing a specific idea. How many tests are acceptable? Well, that depends. Obviously, if you plan to test all possible combinations of variables then there is no specific prediction and this is not a very severe test. Common recommendations are related to degrees of freedom, such as do not do more comparisons than the number of conditions minus one. That is, if you have three conditions, plan two pairwise comparisons. Another technical alternative is to make only orthogonal comparisons, which roughly equates to only testing the data from any given condition at most once. Whilst both of these approaches are cautious, they are rarely of practical use as they omit important comparisons and, in particular, in exploratory analysis, it is simply not known in advance what are the interesting comparisons.

13.3.3 The Bonferroni Correction

The other most common technique is called the Bonferroni correction. In this approach, the reality of needing multiple tests is acknowledged but to protect against interpreting p-values too liberally, the level of significance is lowered. Typically the standard $\alpha = 0.05$ is divided by the number of tests done.

A Bonferroni correction therefore guarantees that Type I errors across the set of tests do not occur with a probability above $\alpha = 0.05$. However, with even a modest set of tests, the level of significance becomes very low for any single test. This corresponds to only indentifying very large effects, which runs the risk of inflating Type II errors: missing effects that are real but not big enough. This contradicts the whole point of an exploratory analysis because huge effects would be obvious in any case without the need for more careful exploration. In general, then, the conservative nature of Bonferroni corrections is too much for a practical exploration. As always, it is better to look at effect sizes rather than p-values and to consider the meaningfulness and explanations behind effects as the outcomes of an exploration.

13.3.4   Bayesian Methods Can Over-Test Too

It is tempting to think that over-testing is a problem of NHST and a fixation on p-values, but it is not really. It is just most apparent in NHST, because Frequentist approaches to probability make it obvious that the more tests that you do the more likely it is to get significance. In the Bayesian context, it is perfectly reasonable to compare any two hypotheses related to a study and use Bayes factors (see Chapter 3) to see which is better supported by the data. However, Bayesian discussions often neglect the importance of the prior probabilities. An exploratory design with multiple independent or dependent variables corresponds to multiple independent hypotheses, so each possible combination of hypotheses being held true or not true must form the minimal set of disjoint hypotheses and which must sum to 1 (Jaynes, 2003). What is the right way to allocate prior probabilities to this huge variety of hypotheses without introducing your own biases into the analysis? And if a uniform prior is used, giving each combination of hypotheses an equal probability then getting an interesting posterior probability is the same problem as a Bonferroni correction: effects would have to be enormous before being declared interesting.

If this discussion is outside of your experience, then here is a different way to think about it. Within the Bayesian framework, there is always a unique hypothesis that says your data should be exactly as you obtained it. This will get maximal support from your data in a Bayesian analysis over and above any other hypothesis. The way that Bayesians avoid fooling themselves is by allocating a *very* low prior belief to this hypothesis. This way, no amount of evidence from an exploratory study should promote the special hypothesis above more sensible hypotheses. This same reasoning should be in play in an exploratory study as well, so that Bayesian analysis does not mislead the researcher into having too much belief in what is essentially an arbitrary idea amongst many other arbitrary ideas.

13.4   Fishing

The move from exploration to fishing is really one of honesty. In fishing, like exploration, the goal is to explore the data to see possible patterns. The difference is that with fishing the exploration stops once an interesting pattern, in particular statistical significance, is found. The problem comes if the result of a fishing expedition is presented within the argument form of a severe test. For example, suppose we have an experiment that is intended to see if a new design of a medical infusion pump reduces the number of errors made by users programming the device. We might find no effect in comparison to the old pump and so of course it is a disappointing result. As good researchers, we explore to see if there are other variables that we have measured that might explain why our good design did not produce the desired results. Suppose we find after a couple of tests with different demographic variables that age is important. Older users made fewer errors with the original device whereas younger users made fewer errors with the new design (an interaction effect). Well, the reasons for this are obvious: older users are more set in their ways and adapted less well to the new design than the more cognitively flexible younger users. It's a good story but it is just that. We had no reason to expect age, out of all possible demographic factors, to be important. But if we now write up the study as finding a significant interaction and one that moreover we anticipated (why else would we have measured age?) we are misleading our readers. And ourselves.

Put like this, fishing seems like blatant deception, which only the most dishonest of researchers, and certainly not you and I, would ever engage in. But really I think the point is that we deceive ourselves as much as anyone else. A null result can be terribly disappointing. All the hard work of designing a good study, of developing the systems that go into it (and even the systems that are tested like the new infusion pump) and the effort of running a good sample of participants: all of this comes to nothing if we get a null result (or equivalently a small effect or a neutral Bayes factor). But we are foolish to ignore data and so it is right to explore to see if there are patterns there that help us to understand such a fruitless use of resources. And suddenly there's a glimmer: within our muddy data a shimmering nugget of significance gold and all is not lost. Moreover, we can think of good reasons why that gold is there. And surely, if we can learn from this experience so can others too. Age is an important influence in infusion-pump error studies. We're not trying to fool anyone: we're just trying to learn from our data. Right?

But in this process, we have lost all sense of a severe test. We most likely have over-tested the data including lots of different variables in our analysis. So Type I errors are potentially abounding. And age was never being severely tested. If it were, we would have sampled much more carefully for age and controlled for the causal implications of age as well, such as the amount of experience users have had with particular pump designs.

Without a severe test, we have nothing except interesting patterns. The problem is that there will always be patterns. From Earth's rather ordinary location in the unfashionable end of the western spiral arm of the galaxy, stars are essentially distributed at random across the night sky. Yet nonetheless we, and countless generations before us, see patterns in the stars: the distinctive ladle-shape of the Great Bear; the W-shape of Cassiopeia; the beautiful Southern Cross; and the surely impossible coincidence of a star that remains fixed at the North Pole of the sky. A random distribution of stars across a night sky will always present patterns and if a U-shape or an L or an 8 is just as interesting a constellation as a W-shape, then the chances of finding something interesting is really quite likely.

Fishing is not avoided by using large datasets either. It is tempting to say that large datasets, by which I mean the many thousands and tens of thousands of participants, who might come from game analytics or social media, represent the population well enough that any pattern is representative of the population as whole. This may be true but remember that correlation is not causation. While there may be apparent connections between variables in a large dataset, it is impossible to know if they are meaningful or robust. To help see this, consider a very modest set of measurements found by gathering the responses to 30 Likert items on a five-point scale. The space of all possible responses to this questionnaire has 5^{30} unique points, roughly 9.3×10^{20}. Any sample of humans (even all of them!) is vanishingly tiny in comparison to such a large number and so there will always be unevenness in the distribution of responses that give the semblance of meaningful patterns.

If you are in any doubt that researchers really do do fishing, there is an interesting analysis of when psychology researchers include covariates in their analysis. There is an indication that across the studies covariates are often included to ameliorate a null result (Lenz and Sahn, 2017). I do not believe that psychology researchers are systematically deceiving the world, but rather they are prey to the pressures of publication, a sociological focus on p-values and a general lack of awareness of the problems of statistical analysis (Chapter 2). Feynman's requirement of radical honesty in science (Feynman, 1985) takes a lot to live up to.

13.5   Some Rules of Exploration

In many situations, then, data contain all sorts of patterns. Some patterns are meaningful results of how the world works. Some are chance. Exploring data brings out these patterns, so it is important that researchers go exploring as there may be gold to be found. But when does exploring become over-testing and, more alarmingly, when does it become fishing. Here are some general rules of thumb that I use to guide my researchers and myself.

Severely test just one idea

An experiment works best as a severe test when there is one key idea that the experiment is designed to attack. It may also test other things but less severely. And the result of having just one idea being tested is that there is only one statistical test in the analysis that addresses it. All other tests are exploratory.

Do not get fixated on significance

Given the problem of Type I errors in over-testing or Type II errors with Bonferroni corrections, probably the best approach to exploratory analysis is to use $p < 0.05$ as a marker of a pattern but always to interpret it as if it is just chance. Unexpected effect sizes, even non-significant ones, might be worth exploring further.

Do not do more statistical tests than participants

You might think this is ludicrous but I regularly see studies where this rule is broken. Just recognise that there is a limit to how much information you can get from your data.

Do not hide exploration

You may have done a lot of tests and found interesting patterns but do not hide the tests that did not give interesting results. They frame the degree of over-testing and if you stick to the other rules then being clear about what you did does not undermine the value of what you found.

Do another study

If you really think you have found something interesting, then devise the study that tests that. That way you will either get more evidence of your idea or you will find that it is not a simple as you thought.

Be honest

I do not think we ever set out to be dishonest but it is so disappointing when a study does not work. No matter how hard that feels, learn what you can from the lack of result and *do another study*. Fishing might help you get published but it will not really help you do good science and I am pretty sure that is what you want to do.

References

Costa Jr., Paul T. and Robert R. McCrae (1992). 'Four ways five factors are basic'. *Personality and Individual Differences* 13.6, pp. 653–665.

Denisova, Alena and Paul Cairns (2015). 'The placebo effect in digital games: Phantom perception of adaptive artificial intelligence'. *Proceedings of the 2015 Annual Symposium on Computer-Human Interaction in Play*. ACM, pp. 23–33.

Feynman, Richard (1985). *Surely Youre Joking, Mr Feynman*. Vintage.

Jaynes, Edwin T. (2003). *Probability Theory: The Logic of Science*. Cambridge University Press.

Lenz, Gabriel and Alexander Sahn (2017). 'Achieving statistical significance with covariates and without transparency'. *Meeting of the Midwest Political Science Association, Chicago, IL*. Available at: www.osf.io/c2qvb.

Royall, Richard M. (1976). 'Current advances in sampling theory: Implications for human observational studies'. *American Journal of Epidemiology* 104.4, pp. 463–474.

CHAPTER 14

When Is a Correlation Not a Correlation?

Questions I am asked:

▷ I've found an *r* value of [*insert value*]: is this a strong correlation?

▷ Why do I need to look at a scatterplot when the correlation is clearly very strong?

▷ Can I use structural equation modelling [*or other fancy correlation-based method*] to analyse my experiment?

Correlation is not causation. This is well understood (Munroe, accessed 2017). When we measure two things about someone, for instance, how immersed players were when playing a game and how many points they scored in the game, these two things can be related to each other. For example, the more immersed a person is in the game, the higher the score. This is correlation. However, correlation does not say why the two measures are related. A person might find that they are getting a good score and this leads them to feel more immersed in playing. Alternatively, they might find themselves immersed in the game and so put in more effort, which leads to a good score. Either of these relationships could be considered causal: as a result of one measure being high, the other was also high. And the third option is that something else was causing both variables to go up: some players found the game controllers particularly suited to their skills so they were more able to become immersed in the game and they were able to get more points. Although there is a relationship between our two measures but they are not causing it; it is being caused by something else that we have not measured. And sometimes, there can be spurious correlations where things correlate and there is no common link between them, such as the divorce rate in Maine, USA and the per capita consumption of margarine.[1]

---

[1] http://www.tylervigen.com/spurious-correlations

What is surprising is that, sometimes, something that looks like a correlation is not only not causal but not even a correlation either. There are various traps in correlations that can catch out the unwary. First then, let's be clear by exactly what we mean by a correlation.

14.1   Defining Correlation

In the world of measuring people and, in particular, people's use or experiences of interactive systems, no relationship between two measures is likely to be perfect. There will be some room for uncertainty: knowing a person's score in the game may give some indication of their level of immersion but not precisely. This is in part because no two people are the same, so their feelings of immersion may differ even if they get the same score. Also, measures of things like user experience are not precise either, so even if two people have the same immersive experience, the measure we use may introduce differences in measured values of immersion. In particular, I tend to use a questionnaire in my research to measure immersive experience (Jennett et al., 2008) and people with the same experiences may well give different answers to the questions.

In order to express the imprecision in correlation between two variables, we use a correlation coefficient. Usually this correlation is Pearson's r, often called *the* correlation coefficient. The range of possible values of r is from -1 to $+1$. The sign indicates whether the correlation is positive, the two values in general go up together, or negative, as one value goes up the other value goes down. The extreme values of $+1$ and -1 mean there is a perfect correlation, so that knowing the value of one variable it is possible to precisely calculate the value of the other. A correlation of 0 of means that there is no correlation and so knowledge of the value of one variable gives no knowledge or constraint on the value of the other variable.

Of course, in practical HCI contexts, no correlation is perfect and real r values are somewhere between -1 and $+1$. Some typical scatterplots for different correlations are given in Figure 14.1.

The key problems arise from the fact that Pearson's r is closely linked to the *linear* relationship between the two variables. A linear relationship is usually described through a regression line. A regression line is a line through a scatterplot that in some sense allows you to predict one variable from another. It matters which way round the variables are so, to be concrete, let's say we want to predict the level of immersion from a player's score in the game. So the score is the predictor variable, typically called x,

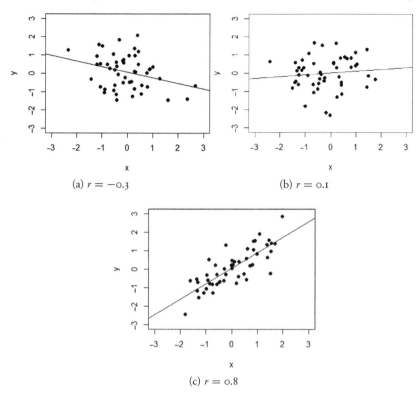

(a) $r = -0.3$ (b) $r = 0.1$

(c) $r = 0.8$

Figure 14.1   Scatterplots with regression lines demonstrating three different correlation
coefficients between two normally distributed variables.

and immersion is the predicted variable, typically called y. Of course, no
prediction is precise, so the predicted immersion is labelled as $\hat{y}$ to show
that it is a possibly inaccurate estimate of some real value y.

A regression line is a line through the actual data that is in one sense
optimal in that, over all players, the average difference between the score
as predicted by the regression line, $\hat{y}$, and the actual score, y, is as small
as it could be. Any other line through the data would give a bigger average
difference between predicted values and real values. I am glossing over what
I mean by average here but in case you are interested the regression line is
actually minimising the root mean square differences of the predicted values
from the real values. As far as this discussion goes, though, it is wrong but
good enough to think of the average difference as the mean of the distances
of the data points from the regression line.

The link between regression and correlation is that the closer a correlation coefficient is to $+1$ (or -1), the closer the real data are to the regression line. A perfect correlation of $+1$ (or -1) means that all the data points lie precisely on the calculated regression line. And a correlation of 0 means that the best prediction of a y value for *any* value of x is simply the mean of all the observed y values: knowing x does not help you predict y values at all. Or equivalently, the regression line is horizontal. This is of course provided that there is a meaningful linear relationship between the variables. Two things can strongly interfere with such a relationship:

- Outlying points
- Clusters in the data

It is also possible to have non-linear correlations where the variables are related but not in a way approximated by a regression line, for example in a curve. But here I am not concerned with these but rather interested in understanding what might make even simple linear relationships difficult to interpret.

14.2   Outlying Points

As with any data where means are calculated, an outlier can have a substantial effect because a point that is a large distance away from the others has a disproportionate effect on the mean (Chapters 8 and 11). With correlations, because the regression line optimises the average distance of data points away from the line, things are bit more complicated. If the outlier would be perfectly predicted by the regression line through the rest of the data, then the outlier has no effect on correlation because the same regression line still fits the data. However, if the outlier is a long way from the regression line that would fit the rest of the data, then the outlier can move that regression line a long way towards the outlier and away from the regression line that is best for the rest of the data.

The effect of outliers on correlations is therefore complex. It is the combination of x and y values, not either value alone, that influences the regression line and hence the correlation. And the outlier's influence can both increase the r value of an otherwise weak correlation or decrease the r value of an otherwise strong correlation, as can be seen in Figure 14.2. Specifically, the leftmost point in both scatterplots corresponds to the player with the lowest score in the game. If that lowest scoring person has a high immmersion then they can substantially reduce what was a

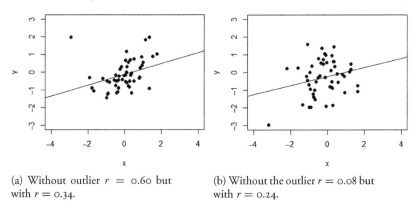

(a) Without outlier $r = 0.60$ but with $r = 0.34$.

(b) Without the outlier $r = 0.08$ but with $r = 0.24$.

Figure 14.2   The weakening and strengthening effect of a single outlier on correlation coefficients of 50 other data points. In each case, the outlier is the leftmost point.

reasonable correlation (Figure 14.2a). Conversely, if they also have a very low immersion then they can make it seem like there is a (modest) correlation where there is essentially none (Figure 14.2b). Notice also that in those figures there are 50 data points other than the single outlier. This is not just a problem for very small samples.

14.3   Clusters

First, what do I mean by a cluster? Informally, a cluster in data is a clump of points in the data that form a distinct group, which sits apart (at least to some extent) from the rest of the data. In some contexts, it is positively useful to find clusters, as these may correspond to particular types of users or particular styles of behaviour of users (Oberski, 2016). However, when it comes to correlations, which is about how variables relate to each other *overall*, then clusters can cause problems.

This is probably best understood by first considering an example. Figure 14.3 shows a scatterplot with two clearly identifiable clusters. In our example, these would correspond to two groups of players, one of which has low scores on the whole and the other has generally high scores. Within each group there is no particular relationship to immersion. Of course, this is exaggerated to illustrate the point but there is an overall modest correlation of 0.3, which is a lot larger than the correlations of the two clusters separately. What is happening is that the two clusters effectively define two points, basically the centres of mass of the clusters. And between

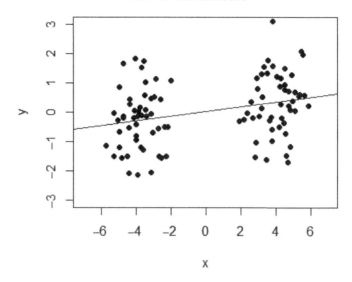

Figure 14.3   The effect of clustering on correlation: the correlation within the left cluster is $r = 0.088$, within the right cluster is $r = 0.17$ and of the two groups together is
$r = 0.30$.

any two points there is a straight line so the further apart the clusters, the more like points they are relative to the overall scale of the picture, and so the stronger the apparent linear relationship between the variables overall. And since correlation is about the strength of linear relationship between variables, clusters tend to suggest stronger correlations. But in fact, within any one cluster, there is only a very weak relationship between the two variables.

One way to try to dismiss this problem might be to say that though each group does not demonstrate the correlation, the two groups together give both a bigger sample and a wider range of values. But if correlation is in part about how much you can predict the y value from a given x value, give me an x value for a particular person and I might confidently tell you which cluster that person belongs to but not their y value. Or, to put it another way, give me a person whose x value is 0, and I have no grounds on which to predict a y value from the data in Figure 14.3.

In practice, of course, clusters are not necessarily obvious and in large datasets, such as survey or questionnaire data, they may be impossible to spot. However, there is one crucial place in HCI where we might expect clusters and that is for experimental conditions. Imagine an experiment of people playing the same game but using two different controllers, say a mouse and keyboard versus a Playstation controller, to see the effect on immersion. Suppose further that the Playstation controller gives more control and hence allows players to get higher scores. Now the two experimental conditions are intended to produce different clusterings of data *by design* (see Chapter 1). But if in the course of the analysis you also wish to see the relationship between score and immersion, then the experimental conditions are confounds in the correlation because they cause clustering in the data. In particular, if there is a significant difference in immersion, as hoped for in the experiment, then even small differences in mean score between the conditions could lead to an artificially high correlation coefficient, just like that shown in Figure 14.3.

14.4   Avoiding Problems

As might already be obvious from the figures presented so far, one way to avoid some problems with correlations is to look at the scatterplots. In fact, to go back to the title of this chapter: when is a correlation not a correlation? When you haven't looked at the scatterplot. Scatterplots are not a failsafe way to spot outliers and clusters, but they are certainly a good place to start.

If you do see an outlier in your data, you are then back at the question of what to do about it (see Chapter 8). But remember, an outlier in data may represent an unusual person in your sample but not necessarily a person unrepresentative of your users. If that is the case (or you cannot show that it is not), an analysis with and without the outlier may be illuminating.

Another alternative is to use a different correlation coefficient. Pearson's r is only one possible correlation coefficient and another that is commonly used is Spearman's ρ (rho, pronounced to rhyme with toe) (Howell, 2016). This is very closely related to Pearson's r but instead of directly correlating the x and y values, the values are first ranked and Pearson's r is calculated for the rank values. This way, for a sample of 100 participants, an outlier can only have a worst ranking of 100 and so not be too far from the 99th participant no matter how extreme the actual value is. While outliers can still affect Spearman's ρ, the effects are not so extreme. The only problem with Spearman's ρ is that it is not so immediately intuitive to interpret but actually in practice, except where there have been outliers, I have never seen any practical difference in the values of r and ρ.

Solving the problems of clusters is not so simple. Spearman's ρ is also affected by clustering in the same way that r is. But at least one lesson is to avoid correlating over experimental conditions. At best, you are going to be re-testing the experimental hypothesis. At worst, you will find misleading relationships between variables.

14.5   A Final Warning

Correlations sneak into statistical methods in all sorts of places and just because you are not looking at an *r* value does not mean that correlations are irrelevant. Structural equation modelling (Beaujean and Morgan, 2016), multivariate regression (Hair et al., 1998), factor analysis (Kline, 2014) and mediation/moderation analysis (Hayes, 2013) all rely on correlations to produce their results. There is an increasing use of these sophisticated techniques in HCI to analyse data, with the result that there are places where false correlations can strongly influence the findings of such studies. As we have seen, this can particularly be a problem where these techniques are used to analyse data across experimental conditions. The further problem is that in such cases, simple scatterplots are not necessarily enough to see the outliers or clusters in the data because the data are often multidimensional. Complex or sophisticated methods may seem to be better but can hide issues that really matter when it comes to interpreting results. I think this is a very good example of where simpler statistics can be better.

References

Beaujean, A. Alexander and Grant B. Morgan (2016). 'Latent variable models'. In: *Modern Statistical Methods for HCI*. Ed. by Judy Robertson and Maurits Kaptein. Springer, pp. 233–250.

Hair, Joseph F., Ralph E. Anderson, Ronald L. Tatham and William C. Black (1998). *Multivariable Data Analysis*. 5th. Prentice-Hall.

Hayes, Andrew F. (2013). *Introduction to Mediation, Moderation and Conditional Process Analysis: a regression-based approach*. The Guildford Press.

Howell, David C. (2016). *Fundamental Statistics for the Behavioral Sciences*. Nelson Education.

Jennett, C., A. L. Cox, P. Cairns, S. Dhoparee, A. Epps, T. Tijs and A. Walton (2008). 'Measuring and defining the experience of immersion in games'. *International Journal of Human Computer Studies* 66.9, pp. 641–661.

Kline, Paul (2014). *An Easy Guide to Factor Analysis*. Routledge.

Munroe, Randall P. (accessed 2017). *Correlation*. xkcd. Available at: https://xkcd.com/552/.

Oberski, Daniel (2016). 'Mixture models: Latent profile and latent class analysis'. In: *Modern Statistical Methods for HCI*. Springer, pp. 275–287.

CHAPTER 15

What Makes a Good Likert Item?

Questions I am asked:

▷ Do I need a mid-point in the responses e.g. "Neither" or "No preference"?

▷ How many response options should a Likert item have?

▷ How should the responses be labelled?

Likert scales are a popular tool in HCI. Building on their success as an easy and effective way to gather information on people's attitudes and their psychology (Kline, 2000, p. 49), they have been adopted in HCI to elicit all sorts of subjective data such as people's acceptance of technology (Venkatesh et al., 2003), perceptions of usability (Brooke, 1996) and, of course, user experiences (Jennett et al., 2008). There are many established usability questionnaires that use Likert scales (Sauro and Lewis, 2012, chap. 8). They are a recommended tool in the HCI practitioners' toolkit (Albert and Tullis, 2013; Lazar, Feng, and Hochheiser, 2010; Rogers, Sharp, and Preece, 2011) and Kaptein, Nass, and Markopoulos (2010) found that around 45% of papers in the ACM CHI 2009 Conference used some form of Likert scale.

Given their ubiquity in modern HCI, new researchers may well reach a point where they want to use or develop a Likert-type scale. The concern here is not how to craft a good questionnaire but how to make a good Likert scale item that could go into a good questionnaire. Assuming, therefore, you have done all the right things to work out what items should go into a questionnaire (Chapter 16), what exactly should a Likert item offer as response options?

As might be expected, there has been a lot of research into this topic in various fields, including psychology, marketing and business. But as also might be expected, there is quite a lot of debate! In addition, the concerns of these fields may not cover all of the concerns we have in HCI; after all, we are not (only) psychologists, and understanding the systems we study can be as important as understanding the people who use them. In this essay, I will try to at least present the arguments around each question so that you can see what the issues are. Fortunately, it does all seem to boil down to some concrete advice but also, as far as HCI research goes, there are some important gaps that we might want to work on. However, because the debate can be quite wide-ranging, it would be useful to get some context and terms sorted out before we wade into the issues.

15.1   Some Important Context

The man who started this all off is Rensis Likert (Likert, 1932) (pronounced Lick-ert). He was interested in measuring Americans' social attitudes and needed something that was quick and effective to deploy, so he proposed four different formats for items in an attitude questionnaire. He found that the now familiar format, see Figure 15.1, worked as well as any other and moreover lent itself to a very simple scoring method. To score the attitude scale, use the 1–5 values of each item and add them (taking direction of the phrasing into account) to give a suitable attitude score. It is probably the simplicity of both the items' construction and their analysis that has led to the successful and widespread adoption of Likert scales.

What Likert was specifying first was the structure of the item as a statement to which people might agree and the *response format* as set of options that indicate the level of agreement. A Likert scale is a measurement tool that is made up of a set of items. Thus, a single item was not intended to be a Likert scale on its own. However, as is seen frequently in HCI, single items of this sort are used as measures, for example, Christophersen and Konradt (2011), but there are some very real problems with treating single items like Likert scales (Carifio and Perla, 2008). To avoid confusion then, we will refer to such items as Likert items, to distinguish them from a Likert scale, which is a set of such items intended to form a measurement

I.   We must strive for loyalty to our country before we can afford to consider world brotherhood.

Strongly Approve (1)	Approve (2)	Undecided (3)	Disapprove (4)	Strongly Disapprove (5)

Figure 15.1   A surprisingly current example item from Likert's 1932 attitude questionnaire.

scale. And simply being made of Likert items is not enough to make a questionnaire a Likert scale: the scale ought to have gone through a suitable process of evaluation for reliability and validity, such as described in Kline (2000), and lead to the measurement of a meaningful concept (Chapter 16).

There are also items that look similar to Likert items but differ in response format; indeed, Likert proposed one himself, where the responses are different descriptions or adjectives that are ordered in terms of expressing increasingly strong views in one direction or another. These are related to semantic differential items that are commonly seen as well, for example one measure of their aesthetics of interactive system uses semantic differential items where a typical item asks a person to rate a system between 'cautious' and 'courageous' (Hassenzahl, 2004). Such semantic differential items pose problems in terms of what it means to sum them into a scale so I will not consider them further here and for our purposes are not Likert items.

I mentioned that there is debate around which format of Likert items is best, but what are people debating about? As with any measurement, there are two primary concerns. Are the different formats of items reliable? And, are they valid? (Kline, 2000).

In this context, reliability comes in two forms. Statistical or internal reliability is the extent to which the items of a scale are consistent with each other and so is sometimes referred to as consistency. It is typically evaluated using Cronbach's α (Chapter 17). To give a concrete example, if ten Likert items belong to a scale so that they all measure the same concept, then the score from the first five items should correlate with the score from the last five items. This is called the split-half correlation, and a good scale should have a good split-half correlation. In fact, this should be true of any division of the scale into halves and Cronbach's α is effectively the average of all the possible split-half correlations (Cronbach, 1951). Thus, it gives a useful measure of the internal consistency of the items overall.

The second form of reliability is that, over time, people answer the same items the same way. This is particularly relevant when viewing the items as measures of attitude or personality, because you would expect these to be stable over time and you would expect people to respond to questionnaires the same way over time. The easiest way to test this is to administer the test twice a suitable time period apart (usually a few weeks) and correlate the scores from the Likert scales. A strong correlation shows good test-retest reliability, also called stability .

Reliability tells us that whatever is being measured is reliably found but not what is being measured. Validity is the notion that what is being measured corresponds to the concept that we say it does (Chapter 16) and this is construct validity which is a cornerstone of any experiment (Cairns, 2016). In the context of studying Likert scales, validity is typically measured by seeing if the scales with different response formats agree with each other, for instance by giving similar mean values with the different items and also that the items correlate with some reference criteria of what the concept means. Nonetheless, as will be seen, a lot of the research around response formats considers primarily reliability, yet it was Cronbach, Mr Reliability himself, who emphasised the importance of validity over reliability (Cronbach, 1950, p. 22).

15.2   Should Items Have a Midpoint?

A midpoint in a Likert item corresponds to the label 'Neutral' or 'Neither agree nor disagree'. The issue is whether, when people select that option, they are really ambivalent about their attitude or simply do not have an opinion. This may seem like splitting hairs but think of a concrete example of the response to a statement 'I like massively multiplayer online role-playing games (MMORPG)'. If you have never played an MMORPG, you may have no opinion so your attitude to such games is at best misleading, whether neutral or not. But some people may mark their answers to such questions using the midpoint when in fact they should leave the item blank or use a 'No opinion' option if there is one. The variations are shown in Figure 15.2.

The confusion of whether a midpoint means a neutral view or no view does seem to be a real issue with respondents offering both

1. I like massively multiplayer online role-playing games (MMORPG)

Strongly Disagree	Disagree	Neither	Agree	Strongly Agree

2. I like massively multiplayer online role-playing games (MMORPG)

Strongly Disagree	Disagree	Agree	Strongly Agree

3. I like massively multiplayer online role-playing games (MMORPG)

Strongly Disagree	Disagree	Agree	Strongly Agree	No Opinion

Figure 15.2   Three formats of Likert item: 1. with midpoint; 2. no midpoint; 3. no midpoint but option to have no opinion

interpretations (Nadler, Weston, and Voyles, 2015). However, another study (O'Muircheartaigh, Krosnick, and Helic, 2001) also suggested that respondents actually meant something different when they provided a neutral response from when they provided a 'Don't know' response. The issue of having a midpoint is therefore about whether this confusion of meanings influences reliability or validity.

One problem of having a midpoint is that of acquiescence bias, whereby respondents might wish to seem to acquiesce to the views of the researcher implicit in the statements presented. A neutral option means respondents can comfortably avoid disagreeing even if they do actually disagree. This has been seen empirically, where respondents were given 4-point (no midpoint) or 5-point items and the scale scores were higher for the 5-point items (Garland, 1991). However, without access to some ground truth about true attitudes, it is hard to know if the score were artificially raised through acquiescence bias in the 5-options format or artificially suppressed through frustration with the 4-options format. Differences in means scores of 4-point and 5-point items were seen in Nadler, Weston, and Voyles (2015) but none significantly different and differences in eight out of the twenty-eight items were lower for the 5-point items. In this study, differences between 4- and 5-point items seem to reflect just the natural variations between different groups of respondents.

Alternatively, if there is no midpoint, it may force people to provide a direction to their agreement that they do not really hold. Again, they could opt to leave out the item or to respond 'No opinion' or they may feel that they should have an opinion and so present a more polarised response than they would otherwise give. This was seen in one study (Weijters, Cabooter, and Schillewaert, 2010) where there were items that were deliberately opposite to each other in meaning (this is a common feature of Likert scales that helps provide increased reliability). Ideally, these items should agree with each other when reversed appropriately, but what was found was that there was increased disagreement between such items when there was no midpoint in the responses. This might be because forcing people to respond in a way that they did not truly feel should lead to essentially introducing an element of randomness into the answers and hence reduce the reliability of scales made up of 4-point items rather than 5-point items. O'Muircheartaigh, Krosnick, and Helic (2001), though, found that having 4- or 5-point items had no effect on the statistical reliability of scales.

Overall then, it seems a midpoint is a meaningful thing but the meaning of the midpoint to participants is not always clear. It does seem to add to the picture and is different from a 'Don't know' option. However, whether 4- or 5-point does not seem to have any practical impact on scales, that is, on a set of items as a whole and a lack of midpoint might be a source of frustration to those who really do have a neutral opinion. In the end, as Cox (1980) suggests, it is probably better to have a midpoint than not.

15.3   How Many Options?

Likert's original items had five options (Likert, 1932) and such 5-point items are still very common. However, a piece of folklore I have regularly been told is that people do not like to use the extreme ends of scales, so seven points or more are better. There are also analytical grounds for expecting longer scales to give more precise and more reliable information (Cox, 1980). At the other extreme, (Kline, 2000) prefers dichotomous (yes/no) scales and early work on analysing response formats suggests that it is enough to have only three response options (Jacoby and Matell, 1971).

Let's first address the folklore. There is little evidence in the literature that people avoid the extremes (Matell and Jacoby, 1972) and certainly in my own questionnaire data (amounting to thousands of respondents over a variety of questionnaires) extreme ends of scales are used a lot.

Turning to reliability, across all the comparisons where scales are compared with responses varying from 2 to 19 (Jacoby and Matell, 1971), 4 to 11 (Leung, 2011), 4 to 7 (Weijters, Cabooter, and Schillewaert, 2010), 3 to 8 (Weng, 2004) and even 2 to 100 (Preston and Colman, 2000), there tends to be little variation in the levels of scale reliability: both internal consistency as measured by Cronbach's α and stability in test-retest situations. Also, they tend to produce similar means and standard deviations (after conversion to comparable scoring systems!), even at the item level (Nadler, Weston, and Voyles, 2015).

The reason for avoiding very few response options seems to be because of concern for validity. The number of points in the item does offer more options and hence offers the respondent a better opportunity to express themselves. This suggests that longer scales lead to more accurate measurement and better validity. But beyond a certain point, there are diminishing returns (Cox, 1980) and very long response sets of nineteen points can increase the time it takes to do the questionnaire (Matell and Jacoby, 1972). Furthermore, at the scale level, any properties of individual items is outweighed by the use of more items to improve the scale overall. For a scale (a set of items) that is essentially valid, adding more response options does little to improve the validity of the measurement scale (Jacoby and Matell, 1971).

Having said that, respondents are human and, just as a lack of midpoint can be frustrating, so too can the lack of opportunity to express what you really mean (Weijters, Cabooter, and Schillewaert, 2010). Respondents also have preferences (Preston and Colman, 2000).

Thus, provided you have a good Likert scale, the actual number of responses in the individual items is not too important for measurement but it seems that 5-point and 7-point scales are pretty good both in terms of expressiveness for respondents and their preferences for number of response options.

One final option of course could be to have complete freedom and use a graphical mark on a line to indicate a continuum of agreement or disagreement. Interestingly, a study into this showed that it makes little difference from 5-, 7- or 11-point response formats and in fact the graphical marks corresponded well to a 5- or 6-point format (McKelvie, 1978). It seems 5 to 7 points are about right no matter how they are presented!

15.4   Label All Options or Just End-Points?

Likert's original scales had a verbal label against each option in the items (Figure 15.1). Obviously, for very long scales such as 19-point scales, such verbal labelling may become an exercise in linguistic subtlety but for our proposed 5 to 7 points, sensibles are pretty much established:

- For 5-points use Strongly disagree, Disagree, Neither, Agree, Strongly agree
- For 7-points use Strongly disagree, Moderately disagree, Slightly disagree, Neither, Slightly agree, Moderately agree, Strongly agree.

But it may be possible to be even more simple and just have labels at the end options of Strongly agree and Strongly disagree and leave the respondent to work out the rest, as in Figure 15.3. This is in fact proposed for simplicity's sake in one HCI text (Albert and Tullis, 2013) and again I have also been told further folklore that end-point anchoring produces better results.

Once again though, it is not as important as it might seem. Whether item options are all-labelled or only end-labelled, the overall scales produce similar means (Dixon, Bobo, and Stevick, 1984) and similar levels of internal reliability though there is increased variation in responses (Weng, 2004). All-labelled options also seem to increase the agreement between

1. I like massively multiplayer online role-playing games (MMORPG)

Strongly Disagree	Disagree	Neither	Agree	Strongly Agree

2. I like massively multiplayer online role-playing games (MMORPG)

Strongly Disagree	1	2	3	4	5	Strongly Agree

Figure 15.3   Two formats of Likert item: 1. fully labelled; 2. end-points labelled.

responses to items that are ask the same thing but are phrased in the opposite way from each other (Weijters, Cabooter, and Schillewaert, 2010).

Overall, it seems that labelling all the options helps respondents to think about exactly what they mean. It therefore reduces unnecessary variation in respondents and, while it does not seem to have an overall effect on measurement in a whole scale, that could be useful for improving the power of analysis. Thus, fully labelled items seem appropriate to reduce respondent uncertainty. This could be particularly important for online or self-administered questionnaires (Weng, 2004).

15.5   The Final Story?

The full story seems to be that Likert items are pretty robust to variations in response format. But to keep the respondents happy and to support powerful analysis, 5 to 7 options in an item, all-labelled seem to be the best. And do not use single items but in fact larger scales to support both good reliability and good validity. How many items make a good scale is another, entirely distinct, issue (Chapters 16 and 17).

Indeed, when I planned to write this, I thought that this would be the full story. But in the course of doing the research for this chapter, I would say there are some real gaps in knowledge when it comes to using Likert scales in HCI.

First, in all of this analysis, the researchers into Likert scales have been thinking about measurement of stable responses such as personality or attitudes. But in HCI, we are often concerned about people's responses to interactive systems, that is, their experiences and perceptions, which understandably are not stable over time as people gain more experience with a particular system. This means we cannot rely on test-retest reliability as a useful measure. And there are good reasons to not rest too much weight on Cronbach's α as a measure of internal consistency (see Chapter 17). What criteria could we use to assess the quality of a questionnaire measuring a person's momentary state and not their persistent attitude?

Furthermore, we are not necessarily interested in just one group of users but possibly a large and diverse group of users, even from different cultures, who may have different biases and attitudes to Likert items themselves. There is very little cross-cultural work on response formats, so we have no idea how important it is when we use Likert items across borders.

Third, sometimes it is not people that are central to our studies but the systems themselves. The usability of a system is not a property of a person but of the system. This stimulus-centred approach had not really been investigated by the time Cox (1980) did his analysis and I am not aware of it being updated since then. What is the right format for the robust assessment of the subjectively experienced properties of a system?

Thinking about moving to modern data-gathering techniques, through texts, emails, mobile platforms and so on, just how robust are Likert items across these formats? There are some indications that the format of scales matters when delivered online (Maeda, 2015). Is that always the case? And what about these other platforms? I would particularly highlight possible problems when there is little screen space for a nicely formatted Likert item. Would seven options be frustrating on a smartphone display?

Thus, while the vanilla approach to delivering questionnaires seems to offer concrete and reassuringly robust guidance, that may not be enough for HCI. There are future research projects here that may be invaluable to the field.

References

Albert, William and Thomas Tullis (2013). *Measuring the User Experience: Collecting, Analyzing, and Presenting Usability Metrics*. Newnes.

Brooke, John (1996). 'SUS-A quick and dirty usability scale'. In: *Usability Evaluation in Industry* 189.194, pp. 4–7.

Cairns, Paul (2016). 'Experimental methods in human-computer interaction'. In: *The Encyclopedia of Human-Computer Interaction*. Ed. by Mads Soegaard and Rikke Friis Dam. 2nd. Interaction Design Foundation. Chap. 34.

Carifio, James and Rocco Perla (2008). 'Resolving the 50-year debate around using and misusing Likert scales'. *Medical Education* 42.12, pp. 1150–1152.

Christophersen, Timo and Udo Konradt (2011). 'Reliability, validity, and sensitivity of a single-item measure of online store usability'. *International Journal of Human-Computer Studies* 69.4, pp. 269–280.

Cox III, Eli P. (1980). 'The optimal number of response alternatives for a scale: A review'. *Journal of Marketing Research*, pp. 407–422.

Cronbach, Lee J. (1950). 'Further evidence on response sets and test design'. *Educational and Psychological Measurement* 10.1, pp. 3–31.

— (1951). 'Coefficient alpha and the internal structure of tests'. In: *Psychometrika* 16.3, pp. 297–334.

Dixon, Paul N., Mackie Bobo and Richard A. Stevick (1984). 'Response differences and preferences for all-category-defined and end-defined Likert formats'. In: *Educational and Psychological Measurement* 44.1, pp. 61–66.

Garland, Ron (1991). 'The mid-point on a rating scale: Is it desirable?'. *Marketing Bulletin* 2.1, pp. 66–70.

Hassenzahl, Marc (2004). 'The interplay of beauty, goodness, and usability in interactive products'. *Human-Computer Interaction* 19.4, pp. 319–349.

Jacoby, Jacob and Michael S. Matell (1971). 'Three-point Likert scales are good enough'. *Journal of Marketing Research* 8.4, pp. 495–500.

Jennett, C., A. L. Cox, P. Cairns, S. Dhoparee, A. Epps, T. Tijs and A. Walton (2008). 'Measuring and defining the experience of immersion in games'. *International Journal of Human Computer Studies* 66.9, pp. 641–661.

Kaptein, Maurits Clemens, Clifford Nass and Panos Markopoulos (2010). 'Powerful and consistent analysis of likert-type rating scales'. *Proceedings of the SIGCHI Conference on Human Factors in Computing Systems.* ACM, pp. 2391–2394.

Kline, Paul (2000). *A Psychometrics Primer.* Free Assn Books.

Lazar, Jonathan, Jinjuan Heidi Feng and Harry Hochheiser (2010). *Research Methods in Human-Computer Interaction.* John Wiley & Sons.

Leung, Shing-On (2011). 'A comparison of psychometric properties and normality in 4-, 5-, 6-, and 11-point Likert scales'. *Journal of Social Service Research* 37.4, pp. 412–421.

Likert, Rensis (1932). 'A technique for the measurement of attitudes'. *Archives of Psychology* 140.

Maeda, Hotaka (2015). 'Response option configuration of online administered Likert scales'. *International Journal of Social Research Methodology* 18.1, pp. 15–26.

Matell, Michael S. and Jacob Jacoby (1972). 'Is there an optimal number of alternatives for Likert-scale items? Effects of testing time and scale properties'. *Journal of Applied Psychology* 56.6, p. 506.

McKelvie, Stuart J. (1978). 'Graphic rating scales – how many categories?'. In: *British Journal of Psychology* 69.2, pp. 185–202.

Nadler, Joel T., Rebecca Weston and Elora C. Voyles (2015). 'Stuck in the middle: the use and interpretation of mid-points in items on questionnaires'. *The Journal of General Psychology* 142.2, pp. 71–89.

O'Muircheartaigh, Colm A., Jon A. Krosnick and Armin Helic (2001). *Middle Alternatives, Acquiescence, and the Quality of Questionnaire Data.* Irving B. Harris Graduate School of Public Policy Studies, University of Chicago.

Preston, Carolyn C. and Andrew M. Colman (2000). 'Optimal number of response categories in rating scales: reliability, validity, discriminating power, and respondent preferences'. *Acta Psychologica* 104.1, pp. 1–15.

Rogers, Yvonne, Helen Sharp and Jenny Preece (2011). *Interaction DesignI*. 3rd. John Wiley and Sons.

Sauro, Jeff and James R. Lewis (2012). *Quantifying the User Experience: Practical Statistics for User Research*. Elsevier.

Venkatesh, Viswanath, Michael G. Morris, Gordon B. Davis and Fred D. Davis (2003). 'User acceptance of information technology: Toward a unified view'. *MIS Quarterly*, pp. 425–478.

Weijters, Bert, Elke Cabooter and Niels Schillewaert (2010). 'The effect of rating scale format on response styles: The number of response categories and response category labels'. *International Journal of Research in Marketing* 27.3, pp. 236–247.

Weng, Li-Jen (2004). 'Impact of the number of response categories and anchor labels on coefficient alpha and test-retest reliability'. *Educational and Psychological Measurement* 64.6, pp. 956–972.

CHAPTER 16

The Meaning of Factors

Questions I am asked:

▷ Is it easy to develop a new questionnaire?
▷ What makes a good questionnaire item?
▷ What exactly is a factor?
▷ I have found these factors in a factor analysis but are they right?

In the early days of HCI, the concern was to make systems work better for people, and this was seen in quite objective terms: for example, better systems are quicker to use, have fewer usability problems and cause people to make fewer mistakes (Newman and Lamming, 1995). As HCI has matured though, it has become more concerned with the experiences that people have (McCarthy and Wright, 2004) but unfortunately this means a move away from the earlier, concrete and objective measures. Indeed, some might argue that objective measures have no place in user experience with all experiences being highly situated and specific to each person (Cairns and Power, 2018). However, there is still a recognition that experiences, no matter how specific to each individual, can be helped or hindered by aspects of design. For instance, in my own field of player experience, we are interested in how altering features of games leads to better experiences of enjoyment, engagement, immersion, presence and so on (Cairns, Cox, and Nordin, 2014). To be able to understand how games work and to think about extending the effectiveness of games into other domains, we need to generalise across such experiences in some meaningful way.

Currently, to tap into these subjective experiences, we can only ask people about their experiences. While there is still room for individuated descriptions of experiences such as might arise in an interview study, questionnaires provide a standardised way to ask and interpret questions. With questionnaires, we are able to meaningfully quantify aspects of subjective experiences and so compare experiences, albeit in a very narrow, focused way, between conditions of an experiment or between designs of a system. Thus, questionnaires abound in HCI and can measure all sorts of subjective experiences: acceptance of technology (Venkatesh et al., 2003); satisfaction with an interactive system (Chin, Diehl, and Norman, 1988); immersion in a game (Jennett et al., 2008); engagement in an information system (O'Brien and Toms, 2010); and so on.

In order for questionnaires to be useful measurement tools, either for scientific enquiry in HCI or for the more practical design of interactive systems, we need to be sure that they are effective measures. Psychometrics has grown up as a branch of psychology and has devoted substantial effort, first, to developing useful questionnaires and second, to seeing how these questionnaires interrelate with a view to improving upon them as our knowledge grows (Kline, 2000). HCI has borrowed much from these methods and, in particular, it is recognised that a key step in producing a good quality questionnaire is to do factor analysis.

There are many books that describe how to do factor analysis, my personal favourite place to start being Kline (2014), and there are some books that also provide substantial mathematical detail for those who feel up to it, for example Chatfield and Collins (1980). But aside from the mechanics of doing a factor analysis of questionnaire data, there is an important question of how doing factor analysis leads us to believe that a questionnaire is valid. In other words, after all the effort of doing a factor anaysis, what exactly do factors mean? They just seem to be a whole bunch of items. Furthermore, it might strike you as odd that many questionnaires set out to provide a measure of a particular aspect of experience, say immersion in digital games, but then describe many factors that constitute the experience (Jennett et al., 2008). So are we measuring one experience or many? In this chapter, I try to explain how factor analysis fits into the development of a useful questionnaire. In addition, I will introduce, in an HCI context, the notion of bifactor analysis, which has growing popularity in psychometrics in part because of its ability to resolve this one-but-many problem.

To illustrate the ideas in this chapter, I use an existing questionnaire, called the Competitive and Cooperative Presence in Gaming (CCPIG, 'sea-pig') questionnaire and specifically the Competitive Presence component of this questionnaire. This is for two reasons: first, it is a questionnaire that I was involved in developing (Hudson and Cairns, 2014); and, second, it illustrates the one-but-many problem very clearly.

16.1   From Concepts to Items

The first step in questionnaire development is to generate the items that will end up constituting the questionnaire. The standard format of items seen in HCI is to use Likert items (Chapter 15), where items are statements about a person's experience of using an interactive system. Respondents score an item by indicating to what extent they agree or disagree with the statement as being applicable to their own experience. There are variations on these formats but Likert items are an accepted standard and with good reason.

Response format aside, the key to each item is, of course, the content of the item: the statement that respondents have to agree or disagree with. What the item is about needs to reflect some aspect of the experience that we are trying to understand and quantify. For social presence in games, we knew that people were able to feel connected to other players through playing a digital game (Poels, de Kort, and Ijsselsteijn, 2007; Sherry et al., 2006). This sense of connection through a mediating technology is commonly referred to as social presence, and, colloquially or informally, is simply the sense of being socially with another person. For instance, a telephone enables social presence because we can talk to each other though physically separated. It does not, however, enable complete social presence as we are not able to see each other's facial expressions and gestures, which might add to the sense of social connection or closeness.

When we set out to develop the CCPIG, social presence had already been recognised as an important constituent of player experiences. There was an existing questionnaire to measure social presence in games, the SPGQ (Kort, IJsselsteijn, and Poels, 2007). However, we felt that this was not sufficiently nuanced to capture the way people may connect to other players. Specifically, the SPGQ was not easily adapted to consider social presence between people that may depend on their roles as either opponents or teammates. To this end, the CCPIG was developed and the competitive component of it was explicitly intended to capture players' sense of social presence with their opponents, which we have simply called competitive social presence.

Underlying this development was our belief that competitive social presence can be quantified, and this is not unreasonable. Some games give little sense of being connected with other players, say in a team death match where opponents are merely targets in a chaotic melee of action, whereas other games, like *Puji*, which due to its configuration of gameplay and controls, can produce a sort of hyper-presence with other players (Hudson and Cairns, 2014a). We are not expecting to be able to precisely quantify social presence (competitive or not) for each individual player in games but rather to capture an indicator of the degree of social presence of all players towards their co-players that a game can offer. With our current state of knowledge, that should be enough to allow us to start to make progress in understanding how games bring people together.

An item that could naively capture competitive social presence might be:

I felt competitive social presence with my opponent

But of course, this requires the respondents to have an in-depth and accurate knowledge of what is meant by competitive social presence. While people may have some ideas, most players are unlikely to have a full picture of what social presence means, let alone what it is in the context of games. Instead, items are intended to capture some concrete aspect of the experience of competitive social presence that respondents do understand. Such aspects may arise from the literature on social presence (Lombard and Ditton, 1997), existing questionnaires used in other contexts (Biocca, Harms, and Burgoon, 2003) and our qualitative research about how players talk about their experiences (Hudson and Cairns, 2014a). Example items are:

- I was aware of the presence of other players
- I felt tense/on edge while playing my opponents
- I felt the need to beat my opponents
- The game was a battle of wits

Hopefully, you can see that the items are looking at how people might feel connected to other players. Each addresses a different aspect but, at the same time, there is a plausible sense of them being about how connected a player feels to the opponents.

Making items is not easy. There are lots of traps in writing good items, not only so that people can understand and answer the items but also so that they do provide a meaningful connection to the concepts of interest (Oppenheim, 2000). Items need to be made with care but, at the same time, there will be plenty of opportunity to refine them. Early on in the process, therefore, it is not possible to have too many items. Generate as many items as you can that you think are relevant.

However, once the initial item pool is generated, the next step is to review, refine and remove items. This is usually done through some form of expert review where people who understand the concepts that are of interest interrogate each item for its relevance to the concepts and the ability of respondents to answer it. Also, overlapping items or items with very similar wording are merged or removed to avoid bloated specifics and the problems they cause (Chapter 17). After this, there is typically a small trial of the questionnaire with one or two respondents who are in the target audience. In the case of the CCPIG, these were gamers, who played a game and then filled out a trial version of the questionnaire before discussing which items they found difficult or confusing.

Once the items have been refined and reduced in this way, the result is a version of the questionnaire that is ready for a validation study. At the initial item pool stage, the competitive social presence component of the CCPIG had twenty-eight items, but after the processes of reviewing and evaluating, these were reduced to fourteen items. The items can be seen in Table 16.1.

16.2   From Items to Factors

The foundation for a factor analysis is the dataset produced when a large number of respondents have completed the questionnaire. What constitutes a large number of respondents is not fixed, but common guidelines are at least a hundred or at least five (or even ten) participants per item (Hair et al., 1998). One way to look at this is to think of each item being its own numerical dimension (admittedly with only five or seven points on it for a typical Likert item). Thus, each item adds a dimension to the dataset. With fourteen items, the responses to the CCPIG competitive social presence component make up a cloud of points in a fourteen-dimensional space. Even with a relatively modest number of questionnaire items and only five points on the Likert items, this space has over six billion different points, that is, possible combinations of answers. What you want and need for a sound analysis is to ensure that typical patterns of responses are well represented in this space of possibilities. So, on the whole, the more respondents the better. For the analysis of the CCPIG, we got around 800 respondents, of which 748 gave complete answers to the items of the competitive social presence component. We felt this was a decent size to get some reasonably sound results. It certainly met all of the standard criteria for sample size that are recommended in the literature.

As the purpose of a factor analysis is to produce factors, it is useful to be clear what a factor is. If we lived in a Platonic universe where there is some ideal reality of which our own understanding of reality is but a shadow, then the concepts underpinning a questionnaire would exist in this ideal world and our questionnaire would be imperfectly capturing these ideal concepts. So for example, there would be an ideal concept of a competitive social presence scale on which our experiences of playing games would map to some specific, well-defined value. The CCPIG items would all (imperfectly) reflect in some small part that ideal value. In this ideal sense, a factor is just such a hidden or latent scale that is influencing the observed values of items in the questionnaire. Because the factor influences several items, the items in the questionnaire correlate with each other to reflect the shared dependence on the underlying factor. Moreover, if we were to aggregate the items that depend on the factor, we would be able to approximate the ideal factor score from the observed items scores. This means that the factor is represented by, for example, the mean of the scores of all the items that relate to that factor.

In reality, though, there is no such ideal and indeed until we develop our questionnaire, it is not clear that there could be such a hidden factor that has any sensible meaning. Moreover, even if it did exist, we do not know which items are strongly influenced by the factor and which are more or less independent of the factor. However, if we behave as if such a thing does exist even if we do not know exactly what it is, then we can identify groups of items that co-vary in particular ways and, with the help of some nice mathematics, propose groups of items that correspond to different factors.

Thus, pragmatically (rather than ideally), a factor is a group of items that relate to each other and, if there are multiple factors in a given questionnaire, we would like each item to belong to one particular factor, that is, each item does not appear in multiple groups. I am deliberately vague here about what it means to relate to each other. At a simple level, it means all the items of a factor co-correlate reasonably well with each other and do not correlate so well with items of a different factor. However, it is rarely so simple.

One important concept in factor analysis is that of a loading. The loading of an item on a factor can be intuitively understood as the correlation of the item with the factor. Thus, a high loading means the item strongly relates to the underlying factor whereas a low loading means it only weakly relates to the factor. Conversely, a factor can be understood as the weighted aggregate of the items, with the loadings reflecting the weights. This gives the same picture: the items with the highest loadings are most related to the factor. Loadings are essential in the interpretation of a factor analysis.

A factor analysis would give the clearest picture if the factor loadings came out so that every item in the questionnaire loaded completely on only one factor with a loading of 1, and not at all on any other factor, a loading of 0. This is called the simple structure. In practice this never happens, if only because of imprecision in the measurement process. Items tend to load at least a little bit on all factors but hopefully they have a strong loading on one particular factor and only weak loadings on the others. This is the practical expectation of what a simple structure looks like.

It is also worth noting one further definition of a factor. Returning to the metaphor that response data from a questionnaire is a cloud in a multidimensional space, a factor defines a unique direction through this space. This is not particularly useful in interpreting a factor analysis but it does explain why there are often geometric and spatial terms used when describing factor analysis.

16.2.1   The Methods of Factor Analysis

How factors are found in questionnaire data depends on the mathematical technique used. These vary considerably. There are purely descriptive techniques, most widely used being Principal Component Analysis (PCA), though some would argue that this does not produce factors (Fabrigar et al., 1999). There are also exploratory factor analysis (EFA) techniques, typically Principal Axis Factoring, which seek solutions with a specific number of factors without a preconceived model of what those factors should be like. By contrast, confirmatory factor analysis (CFA) techniques propose a model that is a set of factors and the items that belong to those factors. The CFA methods look to see how well a specific model of factors fits the data. One further step in CFA is structural equation modelling, which not only proposes a model of which items belong to which factor but also how the factors influence each other. Within each of these approaches to factor analysis, there can be a wide range of mathematical techniques depending on the type of data, assumptions on the distributions of the data and, particularly for questionnaire data, whether the individual item scores are treated as continuous scales or discrete, ordered categories.

Most of the techniques for factor analysis (of whatever variety) found in packages like SPSS and SAS are based on Classical Test Theory (CTT). In this paradigm, assumptions are made on how items relate to factors, even though what those factors are is still unknown. Furthermore, CTT tends to gloss over the fact that the Likert items of questionnaires give discrete values that are essentially (but not entirely) ordinal (Chapter 18). Such assumptions were necessary to make the mathematics of factor analysis feasible. However, with increased computing power and advances in the theory of psychometrics, the new approach of Item Response Theory (IRT) is beginning to gain ground in psychology and other domains (Embretson and Reise, 2013). In this theory, the assumptions on items are much weaker and the data are treated as arising from a discrete, ordered set of response values, which is precisely what Likert items are. Thus, this approach is put forward as an improvement on traditional exploratory and confirmatory techniques and can be substituted into either analysis approach. It is therefore IRT that I now use, though, as discussed I have not always used this technique for the simple reasons that it was not available through standard statistical packages until more recently (Chalmers, 2012) and I that did not learn about it until even more recently!

With our context here of developing questionnaires, we logically ought to use an exploratory rather than a confirmatory approach because we simply do not know how many factors to expect and which items relate to them. CFA should be reserved until there is some clearer conception of how the questionnaire should work. Though PCA is widely used to do EFA, there are significant objections to saying that PCA produces factors. This is primarily because factors are intended to be robust constructs that reflect substantive theories (usually in psychology), whereas PCA is just specific descriptions of the data (Fabrigar et al., 1999). EFA is intended to produce factors that are likely to generalise to other contexts. Indeed, IRT starts from a position of theoretical factors further supporting its relevance for EFA.

Having said that, there is, as ever in statistics, some debate. First, it is well acknowledged that often PCA and EFA produce essentially the same results (Fabrigar et al., 1999; Galbraith et al., 2002). This is not always the case and there are both analytical and empirical evidence why EFA should give better results (Widaman, 1993) but in my experience, the differences seen are rarely of material importance. Confusion amongst researchers using these techniques is also compounded because statistics packages, in particular SPSS, implicitly conflate the two approaches by having them as options within the same analysis package.

It should be noted, though, that to do, for example Principal Axis Factoring requires estimating parameters needed in the analysis model, and these of course are estimated based on the data obtained. So there is is a risk that EFA produces better models of a given dataset because it has enhanced the model based on the data, which is somewhat circular. Some even argue that because of the complexity and the risk of over-fitting the data, factor analysis is not worth the effort (Chatfield and Collins, 1980) and PCA is at least an explicit and conceptually clearer method.

However, I would not go so far. It does seem that EFA probably is the more precise technique, particularly if IRT is used, and we should use it, but I would also side with Galbraith et al. (2002) and suggest doing both

PCA and EFA. The principle behind this is to let the data speak.[1] Where there is a real and meaningful structure in data, any reasonable method should reveal that structure. Yes, it may be possible to be more precise or more accurate in the details, but the broad brush picture ought to be the same. Thus, to some extent, it should not matter which technique is used, whether PCA or EFA and using CTT or IRT. If they give more or less the same result, then all is well. If they do not, then there is a problem and it is likely that your data are not suitable in some way. In particular, if different techniques give radically different results, what are you really finding? But I am willing to guess that, in most practical contexts, the techniques will give broadly similar results and you will be able to listen to what your data have to say.

16.2.2   Finding Factors

Having decided on the method of analysis, there is essentially one further decision, which is to decide on the number of factors. This is not a simple decision but there are well established tools that are used. The most common technique is to use what is called the scree plot from a PCA and look for the 'kink' in the plot, see Figure 16.1 for the scree plot of the competitive social presence items.

Technically, the scree plot is the plot of the eigenvalues of the principal components in the order in which they are extracted. A principal component is like a factor but is essentially just a geometric description of the data. You can think of components as initial prototypes for factors. Practically, each eigenvalue is an indicator of the amount of variance captured by each component, where a variance of 1 indicates about the same amount of variance as a single item in the questionnaire. The kink for the competitive social presence questionnaire is pretty clearly between the second and third components. This suggests that we should should be looking for two- or three-factor solutions in our data.

Often the kink is not so clear to see, and a technique called parallel analysis can be used to suggest the number of factors. This basically compares the actual scree plot with a plot generated by a comparable, fictitious dataset but which has been constructed so that there are no factors. Where the two plots cross, that gives an indication of how many factors to extract. The argument for this is that the crossing point indicates where

---

[1] Thanks to Prof. Mark Daley, Western Univesity, for opening my eyes to this principle.

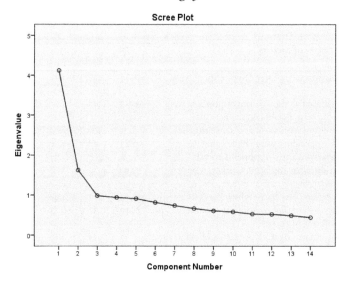

Scree Plot

Figure 16.1 Scree plot of the competitive social presence component of the CCPIG

the data with underlying factors begin to look different from similar data without factors.

In practice, you always have to iterate by looking at solutions around the proposed number of factors and the quality of the factors extracted. So although parallel analysis seems more precise, it is still only an indicator and iterative exploration is needed around the region of the kink (which naturally tends to coincide with the results of parallel analysis).

Having decided on the number of factors to try, run the analysis method in your favourite statistics package. This is definitely a case where doing it by hand is not to be recommended, and indeed could greatly limit your ability to explore your data. In a typical analysis, the first step is to find a suitable solution that fits to the data with the specified number of factors. The next step is to 'rotate' this solution (remember the geometric interpretation of factors). This step is intended to take the set of factors identified and provide a description of them that is as close to a simple structure as possible. As a result of rotation, it is possible that factors correlate with each other, which can make a lot of sense, as we'll see.

The result of a factor analysis is a set of loadings. I explored a one-, two- and three-factor solution for the competitive social presence data using the `mirt` package in R (Chalmers, 2012) to do the EFA. Also, I have used Direct Oblimin to rotate the factor solutions. This is again a

Table 16.1. *Loadings of items on factors for different factor solutions*

No.	Number of factors	2		3		
		F1	F2	F1	F2	F3
1	The presence of my opponents motivated me	−0.44	0.23	−0.39	0.26	−0.12
2	It seemed as though my opponents were acting with awareness my actions	−0.55	0.18	−0.60	0.15	0.19
3	My opponents created a sense urgency	−0.71	0.05	−0.72	0.04	−0.02
4	My opponents were challenging	−0.83	−0.17	−0.80	−0.18	−0.06
5	I acted with my opponents mind	−0.08	0.67	−0.07	0.66	0.05
6	I knew what my opponents were trying to achieve	0.10	0.54	0.06	0.53	0.21
7	The game was battle of wits	−0.43	0.27	−0.48	0.25	0.18
8	I was aware that my opponents might work out my goals	0.49	0.32	−0.31	0.30	0.19
9	The actions of my opponents affected the way I played	−0.18	0.59	−0.16	0.61	0.00
10	I felt I affected my opponents actions	−0.05	0.57	−0.08	0.56	0.10
11	The game was battle of skill	−0.30	0.38	−0.30	0.27	0.02
12	I reacted to my opponents actions	0.11	0.76	0.15	0.79	−0.12
13	I felt tense while playing my opponents	−0.55	0.07	−0.48	0.11	−0.35
14	My opponents played a significant role in my experience the game	−0.44	0.31	−0.36	0.37	−0.28

standard technique used to find a simple structure. The loadings are listed in Table 16.1.

At last, we have factors. But what do they mean? The first iteration is to see which offers the best solution. The general guide to this is to see which items load best on which factor. Treat loadings as correlations, that is they lie between −1 and 1. Close to 0 means very little link between the item and the factor and the closer a loading is to 1 or −1, the more the item 'belongs' to that factor. As a rule of thumb, 0.3 is taken as a cut-off below which an item is not loading strongly on a factor. Note that the sign (positive or negative) of a loading simply indicates the direction of the relationship between the item and the factor but as the factor is simply a mathematical relationship, it is more whether items have the same or a different sign that is important. Positive or negative wording of items on the same factor usually results in the corresponding loadings having a different sign. If all the items on a factor have the same sign, this means that they all work in the same direction and the factor could be 'flipped' to change all the signs round without a change in meaning of the factor.

The two-factor solution in Table 16.1 looks pretty good, as initially suggested by the scree plot. Most items load strongly on one factor and weakly on the other. That is, the solution approximates a simple structure. There are a couple of exceptions: Items 8, 11 and 14 cross-load on both factors above the 0.3 threshold though Item 11 is weakly loading on both factors. And Items 1 and 7 have some degree of cross-loading on both factors even if one loading does not meet the threshold. The three-factor solution looks remarkably like the two-factor solution except one item, Item 13, which seems to constitute the strongest item on Factor 3 and even then cross-loads with Factor 1! This suggests that Factor 3 is not a great factor, as it does not pull items together. Hence the three-factor solution as a whole adds little beyond the two-factor solution. It is also encouraging that a standard PCA with Direct Oblimin rotation produces a very similar two-factor solution with all loadings looking very similar, including the cross-loadings.

Though the two-factor solution is looking promising mathematically, the meaning of the factors is still important and incoherent factors could, and should, result in rejecting a mathematically good solution. To decide the meaning of the factors, you just have to make a decision about which items belong to which factor and try to come up with a meaning for those items collectively.

In our original analysis we used PCA and so based on the slightly different loadings, Factor 1 was decided to be Items 1, 2, 3, 4, 7, 11, 13 and 14. Factor 2 is the remaining Items 5, 6, 8, 9, 10, 12. This perhaps glosses over the fact that Item 11 only weakly loads on either factor but we felt that it was a relevant item, slightly different in meaning from Item 7 and therefore worth keeping in the questionnaire as a whole. Our interpretation of this grouping is that Factor 1 represents a person's experience of Engagement with opponents whereas Factor 2 is more about Awareness of opponents. Of course, the two factors relate somewhat: it is hard to engage with an opponent if you are unaware of them. This relationship is reflected in the fact that after rotation Factor 1 and Factor 2 have a modest correlation of $r = -0.38$. (Note that the negative relationship is just an artefact, because Factor 1 only has negative loadings and Factor 2 only has positive loadings with their constituent items). Also, the IRT loadings presented in Table 16.1 suggest that Items 8 and 11 could arguably belong to the other factor and this would be a source of further correlation if we assign them to the 'wrong' factor. However, there are no hard and fast rules here and, as you see here, different methods result in slightly different results. We have no ground truth to tell us what is right, so we must be satisfied with trying to make the best interpretation that we can acknowledging that other interpretations are possible. Our actual interpretation of the CCPIG is historically situated in the best techniques we had available to us at that time (Cairns and Power, 2018).

This, then, is the basic idea of doing a factor analysis. It is an iterative process of choosing different numbers of factors, conducting the analysis which then leads to partitioning the items of the questionnaire into separate groups. The groups are interpreted for conceptual coherence recognising that there may be overlapping concepts between the groups. From our particular analysis, we decided that Competitive social presence seems to split up into two distinct factors, Engagement and Awareness, that relate to each other in a reasonable way.

16.3   From Factors to Concepts?

The goal of questionnaire development of this sort is to move from a qualitative understanding of a psychological concept to an instrument, the questionnaire, that can meaningfully quantify and measure that concept. However, factor analysis provides a strange outcome. Starting from what was essentially a single concept, a factor analysis often produces distinct factors and hence multiple concepts. So how can that be? Resolving these issues is as old as factor analysis itself, from when Spearman proposed a general factor of intelligence, called g, in 1904 but which can be argued to in fact reflect modest correlations between multiple but distinct factors of intelligence such as mathematical and verbal abilities. Gould (1996) gives an extremely clear account of the historical debate around the meaning of intelligence, g and IQ including, in my view, one of the most accessible descriptions of factor analysis.

In HCI, we hopefully may have more modest goals of pinning down particular aspects of user experiences but the same challenge remains for any analysis. Does a questionnaire represent one concept or several? There is not a definitive answer to this, particularly while we are in the process of defining and constructing the concepts, but some further analysis can help. If there is only one concept, then a one-factor solution should have some merit.

A one-factor solution for competitive social presence is given in Table 16.2. Looking at this solution, it is notable that almost all items load appreciably on the single factor with the exception of Item 6, which even then would meet the 0.3 loading threshold. It does seem that in some sense all the items hang together. This is just using judgement based on loadings, but this can be backed up using a measure of reliability. We cannot use Cronbach α because the possibility of two factors can artificially inflate this (Chapter 17). However, the statistic ω (Omega) provides a measure of reliability when this might the case (Dunn, Baguley, and Brunsden, 2014). This gives $\omega = 0.81$ for competitive social presence so, in fact, the scale does have a reasonable reliability as whole.

A further technique to use is called bifactor analysis (Reise, 2012). In this method, the data are analysed against a specific model where there is both a single generic factor, usually called g, and distinct orthogonal factors based on what the specific, separate factors are expected to be. The term bifactor arises because each item is required to load on two, and only two, factors: the general factor and its specific factor. Thus, bifactor analysis is confirmatory and used to see if the expected factors (such as those described in the previous section) can be better understood as variations on a single factor model. Just such a model was explored using the `mirt` package and the resulting loadings are given in Table 16.2.

Table 16.2. *Loadings of items on the one-factor and bifactor models of competitive social presence with two specific factors, Eng = Engagement and Aware = Awareness.*

No.	Models	1-f F1	Bifactor g	Eng	Aware
1	The presence of my opponents motivated me	0.57	0.44	0.35	0
2	It seemed as though my opponents were acting with awareness my actions	0.62	0.48	0.42	0
3	My opponents created sense urgency	0.65	0.40	0.63	0
4	My opponents were challenging	0.55	0.28	0.68	0
5	I acted with my opponents mind	0.55	0.60	0	0.35
6	I knew what my opponents were trying to achieve	0.32	0.40	0	0.23
7	The game was battle of wits	0.59	0.51	0.31	0
8	I was aware that my opponents might work out my goals	0.49	0.56	0	−0.11
9	The actions of my opponents affected the way I played	0.60	0.69	0	0.14
10	I felt I affected my opponents actions	0.49	0.60	0	0.10
11	The game was battle skill	0.48	0.44	0.20	0
12	I reacted my opponents actions	0.46	0.52	0	0.69
13	I felt tense while playing my opponents	0.53	0.34	0.48	0
14	My opponents played significant role in my experience the game	0.63	0.53	0.33	0

Interpreting the loadings in the bifactor solution, it is clear that g does pick up most of the items with quite good loadings, though on the whole not as high as the one-factor solution. Furthermore, looking at the specific factors of Engagement and Awareness, only three items in Engagement and one item in Awareness load more strongly on the specific factors than on g. This suggests that g is accounting for most of the structure in the data, with the two specific factors adding only a modest amount to the overall model. This informal analysis can be supported by variations on ω (Reise, Bonifay, and Haviland, 2013), which suggests that the general factor of competitive social presence accounts for most of the variance (around 80%) in the bifactor model.

Overall, the one-factor and bifactor analysis suggest that there is a decent, unified, meaningful, general concept of competitive social presence but that there are further distinct aspects due to the Engagement that players feel towards their opponents and their Awareness of their opponents. These separate aspects could be used to provide nuance and subtley in the interpretation of the underlying concept of competitive social presence.

16.4   What Does It Mean?

Questionnaire development is therefore a process of moving from a qualitative understanding of a concept to a questionnaire that can measure that concept, at least to some degree, and can even identify constituent aspects of the concept. Even all this analysis, though, does not imply that there is indeed a meaningful concept that is being measured by the questionnaire. At the very least, we can say there is some coherence to the description of the data gathered by the questionnaire, and that there is some plausibility that this coherence will persist whenever we use the questionnaire. But factor analysis alone does not imply the questionnaire is valid.

To make this more concrete, despite all the work done to develop the CCPIG, first and foremost we do not know what exactly the questionnaire is measuring. We would like to say that it is measuring competitive social presence based on the way in which the items of the questionnaire were devised and selected, but we cannot be sure of that on the basis of the factor analysis alone. Second, it may yet be that the concept of competitive social presence is not real in any robust and useful way (Hacking, 1983), that is, it allows us neither to accurately represent the world of player experiences nor to reliably intervene to influence it.

Factor analysis offers the temptation to reify a concept. We might claim that not only is the CCPIG measuring competitive social presence but it reifies it by giving it a concrete, objective meaning: competitive social presence is what the CCPIG says it is. But the reification of concepts through factor analysis is controversial, particularly when the concept has political value (Gould, 1996). In the world of psychometrics, where the concepts of interest are the robust, stable attributes of individuals like attitudes, personalities and intellectual skills, the meaning of factors is challenging because there is little opportunity to isolate these attributes from the individuals who possess them. However, in HCI, we may actually be in a better position to isolate factors from people. For user experiences at least, these are things that we might potentially influence through experiments. A person's experience should be observed to change with different systems, in different tasks and over time (unlike their personality). And if it is possible to make specific claims about how the concepts involved in user experiences should change then there is the opportunity to put them under severe tests (Chapter 1). Factor analysis is only the start of defining the meaning of concepts and being a little bit clearer about what those concepts are. Experiments are how the concepts are really revealed.

References

Biocca, Frank, Chad Harms and Judee K. Burgoon (2003). 'Toward a more robust theory and measure of social presence: Review and suggested criteria'. *Presence* 12.5, pp. 456–480.

Cairns, A. Cox and A. I. Nordin (2014). 'Immersion in digital games: Review of gaming experience research'. *Handbook of Digital Games*. IEEE/John Wiley and Sons, pp. 339–361.

Cairns, Paul and Christopher Power (2018). 'Measuring experiences'. *New Directions in Third Wave Human-Computer Interaction*. Vol. 2. Springer, (forthcoming).

Chalmers, R. Philip et al. (2012). 'mirt: A multidimensional item response theory package for the R environment'. *Journal of Statistical Software* 48.6, pp. 1–29.

Chatfield, Christopher and Alexander J. Collins (1980). *Introduction to Multivariate Analysis*. Springer.

Chin, John P., Virginia A. Diehl and Kent L. Norman (1988). 'Development of an instrument measuring user satisfaction of the human-computer interface'. *Proceedings of the SIGCHI Conference on Human Factors in Computing Systems*. ACM, pp. 213–218.

Dunn, Thomas J., Thom Baguley and Vivienne Brunsden (2014). 'From alpha to omega: A practical solution to the pervasive problem of internal consistency estimation'. *British Journal of Psychology* 105.3, pp. 399–412.

Embretson, Susan E. and Steven P. Reise (2013). *Item Response Theory*. Psychology Press.

Fabrigar, Leandre R., Duane T. Wegener, Robert C. MacCallum and Erin J. Strahan (1999). 'Evaluating the use of exploratory factor analysis in psychological research'. *Psychological Methods* 4.3, p. 272.

Galbraith, J. I., Irini Moustaki, David J. Bartholomew and Fiona Steele (2002). *The Analysis and Interpretation of Multivariate Data for Social Scientists*. Chapman and Hall/CRC Press.

Gould, Stephen Jay (1996). *The Mismeasure of Man*. WW Norton & Company.

Hacking, Ian (1983). *Representing and Intervening: Introductory Topics in the Philosophy of Natural Science*. Cambridge University Press.

Hair, Joseph F., Ralph E. Anderson, Ronald L. Tatham and William C. Black (1998). *Multivariable Data Analysis*. 5th. Prentice-Hall.

Hudson, M. and Cairns (2014a). 'Interrogating social presence in games with experiential vignettes'. *Entertainment Computing* 5.2, pp. 101–114.

(2014b). 'Measuring social presence in team based digital games'. In: *Interacting with Presence*, pp. 83–101.

Jennett, C., A. L. Cox, P. Cairns, S. Dhoparee, A. Epps, T. Tijs and A. Walton (2008). 'Measuring and defining the experience of immersion in games'. *International Journal of Human Computer Studies* 66.9, pp. 641–661.

Kline, Paul (2000). *A Psychometrics Primer*. Free Assn Books.

(2014). *An Easy Guide to Factor Analysis*. Routledge.

Kort, Yvonne A. W., Wijnand IJsselsteijn and Karolien Poels (2007). 'Digital games as social presence technology: Development of the social presence in gaming questionnaire (SPGQ)'. *Presence 2007*. pp. 25–27.

Lombard, Matthew and Theresa Ditton (1997). 'At the heart of it all: the concept of presence'. *Journal of Computer-Mediated Communication* 3.2, published online.

McCarthy, John and Peter Wright (2004). 'Technology as experience'. *Interactions* 11.5, pp. 42–43.

Newman, William M. and Michael G. Lamming (1995). *Interactive System Design*. Addison-Wesley.

O'Brien, Heather L. and Elaine G. Toms (2010). 'The development and evaluation of a survey to measure user engagement'. *Journal of the American Society for Information Science and Technology* 61.1, pp. 50–69.

Oppenheim, Abraham Naftali (2000). *Questionnaire Design, Interviewing and Attitude Measurement*. Bloomsbury Publishing.

Poels, Karolien, Yvonne de Kort and Wijnand Ijsselsteijn (2007). 'It is always a lot of fun!: Exploring dimensions of digital game experience using focus group methodology'. *Proceedings of the 2007 Conference on Future Play*. ACM, pp. 83–89.

Reise, Steven P. (2012). 'The rediscovery of bifactor measurement models'. *Multivariate Behavioral Research* 47.5, pp. 667–696.

Reise, Steven P., Wes E. Bonifay and Mark G. Haviland (2013). 'Scoring and modeling psychological measures in the presence of multidimensionality'. *Journal of Personality Assessment* 95.2, pp. 129–140.

Sherry, John L., Kristen Lucas, Bradley S. Greenberg and Ken Lachlan (2006). 'Video game uses and gratifications as predictors of use and game preference'. *Playing Video Games: Motives, Responses, and Consequences* 24, pp. 213–224.

Venkatesh, Viswanath, Michael G. Morris, Gordon B. Davis and Fred D. Davis (2003). 'User acceptance of information technology: Toward a unified view'. *MIS Quarterly*, pp. 425–478.

Widaman, Keith F. (1993). 'Common factor analysis versus principal component analysis: Differential bias in representing model parameters?' *Multivariate Behavioral Research* 28.3, pp. 263–311.

CHAPTER 17

Unreliable Reliability: The Problem
of Cronbach's Alpha

Questions I am asked:

▷ What does Cronbach α mean?

▷ How do I know if a questionnaire is any good?

▷ The Cronbach α on this questionnaire has come out below 0.7. Does that mean I cannot use the questionnaire?

▷ Do I really need my questionnaire to have a Cronbach α above 0.9 to be valid?

▷ Should I try to make my questionnaire as short as possible so that participants do not get bored?

It is well understood that if we use a single item or question to measure a complex, subjective concept, like user's trust in a system or competitive social presence in a game (Chapter 16), then we are likely to measure something else unrelated to the concept alongside, such as people's understanding of the particular wording used. To avoid measuring spurious, unknown concepts, the recommendation is always to use multiple items that each differently address the core concept and to aggregate the response from those items into a single score or scale. Put simply, this is because if the same thing is measured with many different items then the error in each measurement should have different sources and so, across a set of items, the errors cancel out and the true concept shines forth. But the question remains, how many items make a good scale? Is four enough, or forty? And what exactly does "good" mean in this context? Central in the development of scales is Cronbach's α (or more simply α in this chapter, where confusion with the threshold for significance should not arise). This is used as a measure of the reliability of a scale. For the budding scale developer, there are clear guidelines as to what is a good α value for a scale: a quick search on Google with the terms 'good Cronbach alpha' throws up several pages that recommend that a good value is 0.65 or 0.7 or above and more than 0.8 or 0.9 is better still. See what you can find. But is that right? Here I want to discuss a bit more the intuitions behind how scales work that lead to a healthy scepticism in interpreting α.

17.1 Reliability and Validity

To make our discussion concrete, we will think about a scale to measure the newly conceived[1] concept of Ensmartness, being the experience of feeling clever as a result of using an interactive system. Any scale we develop would need to be made of individual items that separately address people's sense of Ensmartness. The sum of scores on each item together produce an Ensmartness score (E_S). Having developed such a scale, there are two immediate concerns. First, does E_S really reflect an individual's experience of Ensmartness? That is, is the scale valid? The second question is, when I measure Ensmartness with E_S, how likely am I to get the same measurement by other techniques? That is, is the scale reliable? There is a nice comparison of validity and reliability to accuracy and precision. Validity is a question of accurately measuring Ensmartness, reliability is a question of precisely measuring Ensmartness and the two can dissociate, as seen in Table 17.1.

Under classical test theory (CTT), for an essentially valid measure of Ensmartness, there is some true score, T, but there is also a degree of error, ϵ, in the scales. So we get that:

$$E_S = T + \epsilon$$

Strictly speaking, each of these terms is a random variable and the equation should have a subscript to indicate that the equation holds for the values of

Table 17.1. *A simplification of the dissociation between validity and reliability of a scale and the quality of the resulting scale*

	Reliable	Unreliable
Valid	Good measure on each individual	On average good but wrong on any individual
Invalid	Consistent but wrong	Nonsense

---

[1] That is, I've just made it up

each individual. However, for the sake of visual simplicity (and common practice) this is not indicated and think of this as a general relationship between a set of measured and true values and ϵ, the error of measurement, is different for each person.

The reliability of E_S is captured by the ratio of variance in T to the variance in E_S. When this value is low (close to zero), most of the variation in E_S is due to the error in measurement ϵ and so the scale is an unreliable measure of T. Conversely, when the ratio is high (close to one), then the amount of error is low and E_S is a close, or reliable, measurement of T. The challenge for scale development is that we simply do not know the true value of Ensmartness, T, because if we did, we would not need the Ensmartness scale. What Cronbach's α does is give an estimate of the ratio of variances, even if we cannot directly know T, and hence an indication of the reliability of the measure. The calculation of α is under the assumption that E_S has some degree of validity, that is, it is related to T as specified in the equation above. If E_S is not related to T but instead to some other concept, Cronbach's α cannot tell.

There is a nice, quite careful discussion of the meaning of α in Dunn, Baguley, and Brunsden (2014), where they show that α, like all statistics, is only meaningful in the context of certain assumptions about how items, individually and collectively, relate to the underyling concept. Furthermore, those assumptions are easily violated and this calls into question the sense in using α in scale development. This is not my particular concern right now, because the challenges to α that I raise in this chapter hold even when the items meet all the necessary assumptions and are, therefore, behaving ideally with regards to the underlying true concept, T.

The usual guidance for what makes a good value of α when developing a scale is 0.7 or above and the higher the better (Wikipedia, accessed February 2017). As far as I can tell tracing back the literature, this seems to have come from Nunnally (1978, p. 230) where he says that in the early stages of trying to measure a concept, this would be sufficient, but that up to 0.8 would be better. Others recommend far higher values (de Vellis, 2003). Let's try to see what that looks like in ideal situations.

17.2 A Simple Model

Although it is possible to mathematically analyse the behaviour of α, it may not help illuminate what can go wrong with α. To this end, I will illustrate my discussion with reference to a simple model of an Ensmartness

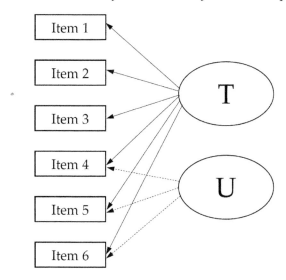

Figure 17.1   An example model for Ensmartness with six items. All items are correlated to a true Ensmartness value, T, and half to the unmeasured factor, U.

questionnaire. In this model, there can be any number of items, all of which correlate with the underlying true value of Ensmartness, T. The correlation coefficient of each of these items with T is the same for all items and is denoted by r_T. In addition, the model includes an unmeasured factor, U, that is, something other than Ensmartness (which will be discussed in more detail shortly). Only half of the items correlate with U but, for those that do, they all have the same correlation coefficient of r_U. A diagram of the model with six items in the questionnaire is given in Figure 17.1.

It is unrealistic that each item correlates with Ensmartness to the same degree, but actually this is effectively one of the assumptions underlying the correct use of α so it is a good assumption for this model. Thus, for r_T values close to 1, each individual item is pretty much measuring the true value, T, on its own. For values of r_T close to 0, each item has almost no relation to the true value.

In the discussion that follows, I will describe the behaviour of the model under varying numbers of items and values of r_T and r_U. I did this analysis by running a simulation of gathering data from a random sample of questionnaire respondents where the model was used to generate the responses to items, which were five-point Likert items. While the mathematical analysis would give precise results, this illustrates that even with precise models, the results of an actual calculation of α is always approximate, due to the natural variation of the samples. Also, the correlation with each item cannot be perfect because the true Ensmartness value is modelled as a continuous scale and each item is modelled as a five-point Likert item. This introduces some variation due to rounding at the very least.

The advantage of using such a model, rather than real data from questionnaires that I have developed, is that I know precisely T (which is not possible in reality) and, through the correlation of T with the Ensmartness Score E_S, what the true reliablity of the Ensmartness scale ought to be. What is important is how α in a simulated sample differs systematically from this true reliability.

Overall then, each run of the simulation consists of 300 respondents to the Ensmartness questionnaire. The simulated respondents' true Ensmartness follows a normal distribution. Their responses to all of the items correlate to the true Ensmartness value and their responses to half the items also correlate with an unmeasured value unrelated to Ensmartness. The unmeasured factor, U, is also normally distributed and apart from this factor, the simulated data meet all the assumptions for an accurate calculation of α.

17.3   When α Is Low

Consider the situation where each item of E_S has only a small correlation with the true value T, say $r_T = 0.3$ and for now we assume $r_U = 0$ so there is no unmeasured factor muddying the waters. While this seems like a very weak link between each item and what we are trying to measure, it is perhaps surprising that even with four items in the scale, the correlation between E_S and T is around about 0.5. This sort of correlation between a pair of variables is actually visible on a scatterplot, suggesting that though weak the questionnaire could start to provide useful information. Note also that such a low correlation corresponds to $\alpha \approx 0.25$ so nowhere near traditional thresholds for reliability. If such a scale could be shown to be valid, then it would provide a quick and dirty measure of the true ensmartness value.

More importantly, even with such a low value of r_T, increasing the number of items has a very useful effect. Though each item is poor, the more items we have, the better the overall questionnaire is. A scale with ten items can give a correlation of around 0.7, which is very respectable, and twenty items up around 0.8. What is happening is that each item is giving a small but distinct bit of information about the true value. Each item also has quite a lot of error, but all the errors are unrelated so they more or less cancel out.

Even with a correlation between E_S and the true value of $r = 0.8$, we get $\alpha \approx 0.64$, which would generally be regarded as inadequate by the traditional guidelines. But actually, we have quite a good scale where we can see clear differences between individuals on Ensmartness and we would be able to see even quite modest systematic changes on Ensmartness between different interactive systems. This is not a trick or accident of a sample that happens to give $\alpha > 0.7$. Even with α at 0.64, E_S is already a sufficiently reliable measure to support systematic investigations. If further it could be shown that this questionnaire is valid, then it would be a more than useful scale to do studies in Ensmartness.

It is also worth noting that when the assumptions underlying α are violated in certain ways, α can be even lower than the true reliability (Raykov, 1997). Thus, a 'low' α is not necessarily an indication of a poor scale. Furthermore, individual items can be very poor ($r_T < 0.2$) provided they do capture some aspect of the true value (have some validity) and you have a lot of them. Though it is probably worth reiterating Nunnally (1978) that if you have forty items in your scale and reliability of 0.3 then it might be worth starting again, because it is a lot of work for participants for at best an uncertain measurement.

17.4   When α Is Too High

It is certainly not uncommon to see really quite high values of α in papers reporting scale development, where high means 0.9 or even 0.95. Not too surprisingly in light of the above, where the scale has twenty items, the items can be moderately related to the true value and collectively give such a high α. I found a value of r_T of about 0.5 gives an alpha of about 0.9; though even with twenty items, to get α of around 0.95 required $r_T \approx 0.65$, making each item on its own a fair measure of the true value, T.

With shorter scales, each item separately needs to be a better and better measure of the true value if the α value is to remain high. For a scale with only six items to have $\alpha = 0.9$ requires *each* item to correlate with the true value at $r_T > 0.7$. For four-item scales, that goes up to $r \approx 0.8$. In such short scales, each item is able to reasonably accurately measure the true value on its own. If the concept is difficult for people to access or articulate, such as Ensmartness, then it sounds a bit too good to be true. And this is not just a theoretical problem: I regularly see scales in the HCI literature of fewer than ten items with such high α values.

Another problem leading to inflated values of α happens when there is more than one concept being measured by the items. For example, when measuring Ensmartness we might also be measuring how happy a person feels to be using an Ensmartening system. Thus, items could both be tapping into the feeling of being smarter and a general feeling of happiness. This is an example of our unmeasured factor, U. The calculation of α cannot account for validity so, even if $r_T = 0$, as r_U increases the apparent reliability goes up. For example, if $r_T = 0$ and $r_U = 0.4$ then with a ten-item questionnaire, α is around 0.4 even though it has no validity whatsoever.

This is of course extreme. But for a scale with ten items and a modest r_T of around 0.3 to 0.4 and a similar level of r_U, the α is inflated above the true reliability of the questionnaire to the point of passing the magic 0.7 threshold and even getting into the 0.8 region of a good reliability. But the scale is not reliably measuring ensmartness; it is measuring two things, only one of which is Ensmartness and the other one, which we have not identified, and that is a real problem if you just want to study Ensmartness.

This situation could arise from a lack of conceptual clarity about what Ensmartness is. Our items are not separating out the concept we are interested in from other concepts that are related to it but distinct. Another way it can arise is if the items themselves are not sufficiently different and correlate because of similarity in wording. For example, 'The system makes me feel smart' and 'I feel smart when I use the system' should elicit very similar responses from respondents regardless of the relationship between either of the items and ensmartness. This is called a *bloated specific*. Instead of each item independently measuring T, the variation in people's responses is correlated because of the wording. If there are several such items, these correlations between items function like an unmeasured factor, U.

Consider the model of a six-item questionnaire where three of the items have very similar wording, so that $r_U = 0.7$. This does not give a perfect correlation between the responses to the three items but nonetheless it has a substantial effect on α. If $r_T = 0.3$ then α is around 0.8, suggesting good reliability. And even with $r_T = 0.1$, α is still frequently above 0.7, suggesting that the six item scale is very reliable, whereas its true reliability for measuring ensmartness is actually more like 0.05. The scale is measuring people's consistency of responses to specific words, not their ensmartness.

There is one further way in which α can be too high, and this is due to a failure to spot clustering. Cronbach's α can be understood as the degree of correlation between the items themselves. And like all correlations, if there is clustering, this can be easily inflated. Clustering can occur if in fact the questionnaire is measuring distinct sets of people. This sometimes happens accidentally but also when a survey is used to gather data in an experiment. The experimental conditions are meant to influence participants and so can lead to clusters in the survey part of the data. Indeed, the true level of reliability can be very low when distinct groups of people (or systems) are being used to establish reliability (Waller, 2008). This is not modelled here, but see Chapter 14 for some examples and discussion about how that happens for pairs of variables. Everything that can cause a problem to a correlation can cause a problem to α.

17.5   Beyond α

Overall, it seems α does not really provide a useful indication of reliability, even in ideal circumstances. It gets worse if the assumptions on the data are not met by real data (Dunn, Baguley, and Brunsden, 2014; Raykov, 1997). There are other measures of reliability that can be used, but they do not necessarily get round the problems illustrated here because these problems are for α under ideal conditions. It is not a violation of assumptions that is causing a problem, but rather that α is not indicating what you might think.

Probably the key lesson is that α, or any measure of reliability, needs to be taken into account alongside lots of other aspects of scale development. First and foremost must be validity. Even if the scale is not hugely reliable, if it seems to capture the concept you are interested in then it might be good enough to get you started in studying that concept. Nunnally (1978)'s reason for wanting $\alpha > 0.7$ is that, in his field of psychometrics, if you have a lower α then you have a lot of error in your measurement and where the field relies on the correlations between variables, as psychometrics does, you do not want to end up correlating with error. However, unlike psychometrics, HCI does not only need to do correlations: it can also do experiments. Even when experiments have to use a low reliability scale, some of the problems can be overcome by controlling for variation between participants through the experimental design. This sort of control is typically not possible in a psychometrics context. For Nunnally, 0.7 was the point at which he could start to do useful work, but even he says that higher reliability is not really of value unless it comes with increased validity.

There is an argument that without reliability you cannot have validity (Crutzen and Peters, 2015) but I would disagree as we saw above when α

can be low but the questionnaire still correlates appreciably with the true value of the underlying concept. You cannot measure concepts precisely with an unreliable scale, but if you have some validity then, on average, your measurements can be accurate and that can be enough to start to do good research and find out useful and interesting things. As your work on the concept develops and your understanding grows, you might, or even should, want to look for a more reliable measure. This is a common process in science (Chang, 2004) though perhaps not one HCI is mature enough to have had much chance to do.

Why then is there an emphasis on very high values of α? This comes from a particular use of psychometric questionnaires, which is for individual diagnosis. If you are going to make a clinical decision on a person, say whether they are depressed and need medication, you want to make sure that the measurement of depression is accurate (Nunnally, 1978). A high α corresponds to a *generally* low level of error and on that basis, a clinician could have confidence that it is a sufficiently accurate measure of any individual. However, in HCI, we are very rarely concerned that individuals reach certain levels of user experience. Rather, we are concerned that, on the whole, experiences improve or change appropriately as the designs of systems change. This does not really need high precision measurement and so looking for very high levels of α is not necessary.

The other lesson is that any assessment of reliability needs to take place in the context of good factor analysis (Dunn, Baguley, and Brunsden, 2014). Where the scale you are assessing is not unidimensional, α does not give a very meaningful value (Sijtsma, 2009). In the model in this chapter there are two factors, T and U, and that is enough to cause α a lot of problems. Fortunately, it is normal to use factor analysis in scale development and α can be a useful measure to look at along the way, including others like item–total correlations and measures of sampling adequacy (Paul Kline, 2000b). In fact, Cronbach (1997) recommended that, when evaluating a scale, it is useful to complete a twenty-nine-point checklist in order assure a sound process. Consideration of internal reliability through α was only one point and the other twenty-eight were just as important.

Finally, avoid small scales. If you have a very short scale, it is probably too specific to be valid (Kline, 2000a). And if the reliability is almost too good to be true then you should be sceptical. In my simulations using the model, which are ideal conditions for calculating α, reliability seems to be achieved consistently for around ten items. But that is the ideal. In practice, longer scales are probably needed to achieve good reliability but too long can be taxing for participants. There's a happy medium to be struck that is best decided by piloting in the context of use. It is HCI, after all.

References

Chang, Hasok (2004). *Inventing Temperature: Measurement and Scientific Progress.* Oxford University Press.

Cronbach, Lee Joseph (1997). *Essentials of Psychological Testing.* 5th. Harper and Row New York.

Crutzen, Rik and Gjalt-Jorn Ygram Peters (2015). 'Scale quality: alpha is an inadequate estimate and factor-analytic evidence is needed first of all'. *Health Psychology Review*, pp. 1–6.

de Vellis, Robert F. (2003). *Scale Development: Theory and Applications.* 2nd. Sage Publications.

Dunn, Thomas J., Thom Baguley and Vivienne Brunsden (2014). 'From alpha to omega: A practical solution to the pervasive problem of internal consistency estimation'. *British Journal of Psychology* 105.3, pp. 399–412.

Kline (2000a). *The Handbook of Psychometric Testing. New York. Routledge.*

Kline, Paul (2000b). *A Psychometrics Primer.* Free Assn Books.

Nunnally, Jum C. (1978). *Psychometric Theory.* 2nd. McGraw-Hill.

Raykov, Tenko (1997). 'Scale reliability, Cronbach's coefficient alpha, and violations of essential tau-equivalence with fixed congeneric components'. *Multivariate Behavioral Research* 32.4, pp. 329–353.

Sijtsma, Klaas (2009). 'On the use, the misuse, and the very limited usefulness of Cronbach's alpha'. *Psychometrika* 74.1, p. 107.

Waller, Niels G. (2008). 'Commingled samples: A neglected source of bias in reliability analysis'. In: *Applied Psychological Measurement.*

Wikipedia (accessed February 2017). *Cronbach's Alpha.* Available at: en.wikipedia. org/wiki/Cronbach's_alpha.

CHAPTER 18

Tests for Questionnaires

Questions I am asked:

▷ What test should I use to analyse a Likert item?
▷ What test should I use to analyse my questionnaire data?
▷ Does it really matter which test I use for questionnaires?

Questionnaires made up of Likert items (Chapter 15), and even individual Likert items, are useful tools for measuring all sorts of aspects of users and their behaviour. In my own field of player experience, questionnaires for quantifying player experiences abound, and indeed there are even several different questionnaires for measuring the same concept (Denisova, Nordin, and Cairns, 2016). Likert items as a way to quantify interactions are also quite common in HCI generally (Kaptein, Nass, and Markopoulos, 2010) and are used to measure cognitive load (Hart, 2006), preference (Kamollimsakul, Petrie, and Power, 2014) and even proposed – to measure usability in its entirety (Christophersen and Konradt, 2011). Given the prevalence and also practical value of questionnaires and Likert items as instruments in HCI research and also the move to improve the quality of analysis of data, the question addressed here is: what is the right way to analyse data gathered by quantitative questionnaires?

Some hold that because Likert-type data are clearly based on ordinal data then they must be analysed using non-parametric methods (Robertson, 2012). Others hold that it is well established that the F-test is robust to substantial deviations from its assumptions and so typical parametric tests are fine (Carifio and Perla, 2008). However, I have also discussed how t-tests and ANOVA are not as robust as sometimes claimed (Chapters 11 and 12) and, for example, the Yuen–Welch as a variant to the t-test is generally preferable to the traditional t-test. And even non-parametric tests can be susceptible to violations from assumptions, leading to misleading interpretations as discussed in Chapter 10. At the same time, Likert items present a very constrained set of values that help guard against the problems found in other, untamed underlying distributions and some of their tameness might also be passed on to questionnaires constructed from them. Also, the newer robust methods do, to a small degree, sacrifice power in order to be more robust and reliable. Is such a sacrifice justifiable in the context of HCI questionnaire data, particularly when datasets might be quite small due to the practicalities of getting a system under development in front of users?

In this chapter, a range of modern statistical tests are examined in relation specifically to questionnaire data. We first consider individual Likert items before turning to data arising from aggregating across a set of Likert items, that is, scores from a questionnaire. The tests considered are:

1. the (vanilla) *t*-test
2. the Welch test, correcting the *t*-test for heterogeneity of variances
3. the Mann–Whitney (MW) test, as the non-parametric version of the *t*-test
4. the Yuen–Welch test (referred in this Chapter as Yuen to keep it distinct from the Welch test), using 20% trimmed means to help resist deviations from normality
5. the Brunner–Munzel (BM) test as test for dominance rather than movement of means
6. the Cliff test as an alternative test for dominance.

The general approach is to take known sets of data, which are either real or constructed to be realistic, and to examine the behaviour of the tests when taking samples from these datasets. As is common practice, the quality of the tests is examined by considering the Type errors and power (Chapter 9). Recall Type I errors arise when a test gives significance but there is no actual difference between the underlying populations, and power is the ability of the test to be significant when there is actually a difference (a Type II error being the failure to see such a difference). It is worth re-emphasising that the focus on Type errors is not because significance is the be-all-and-end-all of statistical analysis but because Type error rates function as measures of the sensitivity and robustness of tests.

18.1   Testing Likert Items

Because of the debate around how to test Likert items, De Winter and Dodou (2010) conducted a simulation to compare whether a *t*-test or a

Table 18.1. *The different distributions (as percentage probabilities) of five-point Likert item data proposed by De Winter and Dodou (2010).*

Type	1	2	3	4	5	Mean	sd
Very strongly disagree	0	1	3	6	90	4.85	0.50
Strongly agree	1	3	6	30	60	4.45	0.82
Agree peak	5	10	20	45	20	3.65	1.07
Agree flat	10	15	20	30	25	3.45	1.29
Neutral to agree	10	20	30	25	15	3.15	1.20
Neutral peak	0	20	50	20	10	3.20	0.88
Neutral flat	15	20	25	20	20	3.10	1.34
Very strongly disagree	80	12	4	3	1	1.33	0.78
Strongly disagree	70	20	6	3	1	1.45	0.82
Disagree flat	25	35	20	15	5	2.40	1.16
Neutral to disagree	10	25	30	20	15	3.05	1.21
Certainly not disagree	1	4	50	30	15	3.54	0.83
Multimodal	15	5	15	25	40	3.70	1.42
Strongly multimodal	45	5	0	5	45	3.00	1.93

Mann–Whitney test was more appropriate for testing data from Likert items. They found that, on the whole, both tests were equally good at keeping the Type I error rate at the level of 0.05, which is where it should be when the level of significance is set at 0.05. They also found that, in general, the two tests were equally powerful except for particular combinations of skewed and multimodal samples, in which case the t-test tended to pick up on the difference in means.

Their work would suggest that overall both tests are equally good; however, they did not consider the more robust forms of the t-test or the tests that are specifically for dominance. Here, I will fill in this gap by repeating their work but with the larger range of tests. The distributions of Likert item data that De Winter and Dodou (2010) used are given in Table 18.1. They cover a good range of typical situations arising from using Likert items in studies.

18.1.1   Type I Analysis

To analyse the Type I behaviour of the different tests over these distributions, for a single trial a pair of samples was taken from the same distribution and then compared using each of the six tests. Being from the same distribution, the two samples should not be significantly different except by chance, that is, in only one out of twenty runs. Both samples were equal in size, and three sample sizes were used: ten, twenty and thirty to give a total of twenty, forty and sixty data points in each trial. These were chosen to be representative of the small samples typically seen in HCI studies. Ten thousand trials were done with each of the sample sizes at a significance level of $\alpha = 0.05$. If the tests were functioning correctly then, over the 10,000 trials, significance should only occur 500 times. To simplify presentation, the number of tests found significant by each test is represented as boxplots in Figure 18.1: the analysis of sample size twenty is omitted here as it consistently gave a picture in between those of size ten and thirty.

What is first seen is that across all the tests and all the distributions, even with small samples, the Type I error rate is rarely above 0.05. The only consistently liberal test is the BM test but even there, the Type I error rate is only slightly raised and below 0.06 (600 erroneous results out of 10,000 trials).

However, the Type I rate dropping too low below 0.05 could also cause concern, as the level of significance influences power. If the Type I error rate is too low, it might mean that the test would not be powerful enough to detect differences when they are there. As Figure 18.1 shows, with sample size of ten, the Type I error rate can be substantially lower than 0.05 for some distributions. In fact, further analysis of the outliers showed that the strongly skewed distributions of *Very strongly agree/disagree* were consistently the outliers seen on the boxplots. This is exactly what was found by De Winter and Dodou (2010) as well.

Basically, where there is strong skew, because there are only a limited number of values in the scale, there is a dominant data value in any small sample. And with very small samples, only one or two data points may be distinct from the dominant value. This effect is made more pronounced by the Yuen test, where trimming of extreme values leaves essentially all data points to be the same dominant value. As a result of this, even for sample sizes of thirty, the Yuen test is *never* incorrect in 10,000 trials of the *Very strongly agree* distribution. This may sound good but very low Type I error rates reflect a lack of sensitivity in the test, which can then appear as a high Type II error rate.

It is noteworthy that for the larger sample size of thirty, the range of Type I error rates with the exception of the Yuen test is very narrow, with all tests showing good behaviour and the non-parametric tests, MW, BM and Cliff, showing the closest adherence to the α of 0.05 across all distributions.

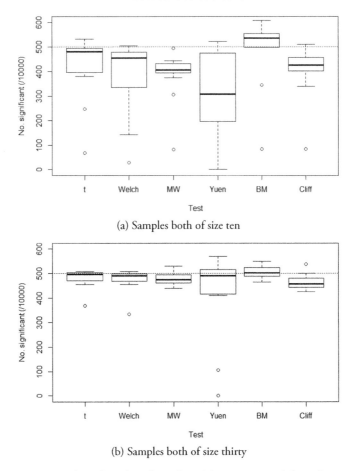

(a) Samples both of size ten

(b) Samples both of size thirty

Figure 18.1   Boxplots of number of significant Type I errors made by each test over 10,000 trials for each Likert distribution and for different sample sizes. The dashed lines are at 500 tests corresponding to $\alpha = 0.05$.

18.1.2   Power Analysis

To test for power, a single trial consisted of samples of the same size taken from different distributions. As there are fourteen distributions being considered, this gives rise to ninety-one distinct pairs from which samples can be taken; again the samples were compared using each of the six tests and 10,000 trials were run. Power of course depends on various factors including the actual difference in the distributions. With perfect power,

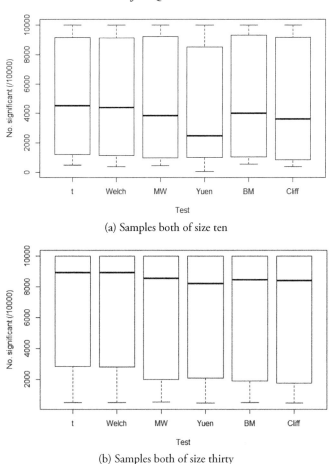

(a) Samples both of size ten

(b) Samples both of size thirty

Figure 18.2   Boxplots of number of significant trials out of 10,000 trials for each pair of distinct Likert distributions and for two different sample sizes.

a test would detect that the distributions were different in 10,000 trials. Also, ideally as samples get larger, power should always increase.

Again to simplify presentation initially, boxplots are used to represent the general distribution of power across the tests, Figure 18.2.

At this coarse level, it is clear that power varies enormously as might be expected because the similarity of the distributions varies substantially. Samples from distributions that are very different are consistently found to be significantly different whereas samples from very similar distributions are rarely found to be significantly different. Also, broadly speaking, the tests are performing comparably, though it is noticeable that the Yuen test has a lower median in both boxplots. Moreover, somewhat more surprisingly, the *t*- and Welch tests have the highest median power and higher lower quartiles for the larger sample size. This suggests that they are a little bit more sensitive than the other tests. This makes sense because it is precisely this sensitivity that makes them problematic in more general contexts.

Encouragingly, all the tests show almost ideal power in distinguishing *(Very) strongly agree* from all the distributions that are predominantly *Disagree* and vice versa.

At this point, it might be tempting to do some statistics to analyse the quality of the statistical tests. Logical circularity aside, this would not be a good idea because the ninety-one pairs are not independent of each other and independence of items is an essential assumption of many tests. Specifically, each distribution listed in Table 18.1 is paired with each of the thirteen other distributions leading to a high degree of dependence between the measures of power for those thirteen pairs. My interpretation therefore takes a more qualitative approach.

A key question is: what are the tests detecting as different? The Likert distributions vary substantially in shape, so there are many ways in which they might be different. However, statistical tests generally look at differences in location or dominance (Chapter 4). From the ideal distributions, it is possible to give an explicit, definitive measures of differences in location and the relative dominance of distributions. Difference in means were used to measure changes in location.[1] The degree of dominance is simply the proportion of scores higher in one distribution than the other. These ideal measures of difference were compared against the power of the best test for a particular pair and the scatterplots are given in Figure 18.3. To help comparison, the difference in means was standardised to be between 0 and 1, with 0 meaning no difference and 1 meaning maximum possible difference. Similarly, relative dominance was standardised to be between 0 and 0.5, with 0 meaning no dominance and 0.5 meaning the maximum dominance of every value in one group being bigger than every value in the other group.

As is to be hoped, as the distributions become more different by either measure, the power of the best test goes up. Also, the larger the sample, the more likely the test is to be significant.

---

[1] Trimmed means were also considered as robust measures of location but they give a very similar picture to the raw means in the constrained world of Likert item distributions so are not reported here.

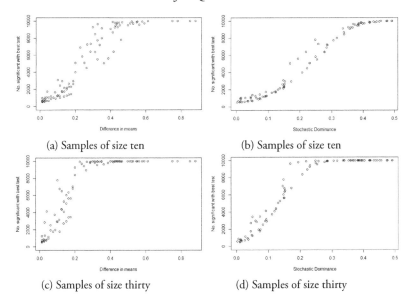

(a) Samples of size ten | (b) Samples of size ten

(c) Samples of size thirty | (d) Samples of size thirty

Figure 18.3   Scatterplots of highest power test against the ideal difference in means or dominance between the distributions.

In terms of what the tests are detecting, it is clear that the level of power of the best test is most closely predicted by the measure of dominance of one distribution over the other. In particular, for samples of size ten, look at the dispersion of power for a difference in means of around 0.3, Figure 18.3a. Power varies between 5,000 significant results to 10,000 significant results for the same difference in location. Thus, in some situations, the best test is very powerful but an interpretation in relation to a change in location would be unreliable. Such a wide dispersion is not seen in relation to dominance.

It is also worth noting that there are some pairs where power is very low, with fewer than 1,000 trials out of the 10,000 producing a significant result, even for samples of size 30. These pairs are mostly those that are around neutral, so *Neutral to agree, Neutral peak, Neutral flat* and *Neutral to disagree*. These pairs also have very similar means (about 3.1) and very little difference in dominance. So it would seem that though these distributions are different they would not reflect a meaningful difference in agreement on a Likert item. The low power is therefore appropriate.

For samples of size ten, the test which most often gives highest power is the BM test, with it giving the highest power of any of the tests for forty-eight of the ninety-one pairs of distributions. This is only distantly followed by the *t*-test giving the best performance in twenty-eight pairs. They only agree on three pairs where they both give the highest possible power of 10,000 significant tests.

For samples of size thirty, the situation is reversed with the *t*-test giving the highest power of the six tests on forty-eight of the pairs and the BM test on forty-one of the pairs, though they agree on twenty-five pairs where they both give the highest power of 10,000. In both sample sizes, the BM test might be performing quite well because it is a little liberal, as seen in the Type I trials, so it will give significance a little more often.

18.1.3   Which Test for Likert Items?

Overall then, it seems there is little to choose between the tests for Likert scales, though the Yuen–Welch test is rather conservative, giving very low Type I error rates but correspondingly lower power. Both the *t*-test and the BM test give good reliability but it does seem that what the tests are best at detecting in this context is the dominance of one distribution over the other. Thus, in terms of accurate interpretation, it would seem that the BM test as a test of dominance is the best test to use on five-point Likert scale data, particularly with small samples.

18.2   Questionnaire Data

Similarly to De Winter and Dodou (2010), Bakker and Wicherts (2014) considered the suitability of different tests for analysing data arising from a full questionnaire, not just individual Likert items. In their second study, they explicitly compared several tests, including all those considered here except for the Welch test. They were particularly interested in the effect of outliers and the quality of the tests. They simulated data based on three types of underlying distribution:

1.   Normal data
2.   Mixed-normal data, which is known to threaten the validity of the *t*-test (Chapter 11)
3.   Simulated questionnaire data, based on a Rasch model, which is a specific version of Item Response Theory
4.   Mixed-normal data for simulated questionnaire data.

Without outliers, the tests performed very similarly with the *t*-test giving a slight power advantage over the others and the Mann–Whitney giving a slight advantage over the Yuen–Welch test. With outliers, of course, the problems with the *t*-test emerged (see Chapter 11). Interestingly, though, within the simulated questionnaire data, the effect of outliers on the *t*-test was much less, though it did have appreciably less power than the other two tests. Again, the Mann–Whitney slightly outperformed the Yuen–Welch. The BM and Cliff tests had performed comparably to the MW.

Bakker and Wicherts (2014) is very thorough work but I am interested in what would happen if the underlying distribution of the questionnaire was not even vaguely normal but came from more extreme distributions. In their first study looking just at the effect of removing outliers on the *t*-test, they had used real datasets that showed considerable skew and non-normality, but they had not carried these into the second study that compared tests.

I therefore simulated questionnaire data based on a simpler model than the Rasch model, namely Classical Test Theory. I was very surprised to find that even with an exponential or a very skewed log-normal distribution underlying the true values that the questionnaire was simulated to measure, the data tended to look remarkably normal. Basically, the process of forcing the value of a latent concept into a set of Likert items tends to lose some of the extremeness in a distribution, with the result that a sum of several Likert items is much closer to a normal distribution than the original distribution. It is quite likely that the simulated data used by Bakker and Wicherts (2014) are quite tame relative to data seen in real questionnaires.

To overcome this limitation, the following trials made use of real datasets from HCI that I have had a hand in either gathering or analysing. The datasets used are based on the Immersive Experience Questionnaire (IEQ) data from Denisova, Nordin, and Cairns (2016), the Competitive and Cooperative Presence in Games (CCPIG) questionnaire which was complemented with a measure of Trust (Hudson and Cairns, 2014) and the User Engagement Scales (UES) (O'Brien and Toms, 2010). Both the overall measures and their subscales are used. For the purposes of this test, they are all standardised to be the average of the Likert items that make up the scales and so range from 1 to 5. Missing values are omitted. These are summarised in Table 18.2.

As can be seen, there is some variation in mean but there is also considerable variation in shape, as seen in Figure 18.4. These therefore represent a good range of realistic datasets for examining the quality of different tests.

Table 18.2. *Summary details of the sixteen scales of realistic questionnaire data.*

Scale	Num. of items	Num. of respondents	Mean	sd
IEQ	31	236	3.32	0.48
IEQ.CI	9	236	3.93	0.60
IEQ.EI	6	236	3.23	0.73
IEQ.RWD	7	236	2.68	0.75
IEQ.Ctrl	5	236	3.33	0.68
IEQ.Ch	4	236	3.20	0.60
UES	31	794	3.54	0.51
UES.FA	7	793	2.71	0.80
UES.PU	8	794	3.98	0.68
UES.AE	5	794	3.68	0.62
UES.RW	11	794	3.68	0.53
CCPIG.Coop	25	753	3.64	0.76
CCPIG.Comp	14	748	3.87	0.56
CCPIG.Eng	8	748	3.69	0.67
CCPIG.Aw	6	748	4.10	0.61
Trust	4	806	3.28	0.96

18.2.1　Type I Analysis

As for the Likert scale analysis, a trial consisted of generating two samples of the same size from the same distribution and comparing them using each of the six tests. There were three sample sizes of ten, twenty and thirty (so twenty, forty and sixty 'participants' in each trial). Ten thousand trials were done with each of the sample sizes. Again, with significance at 0.05, around 500 trials should be significant if a test is working as it should. The boxplots of the distribution of Type I errors is shown in Figure 18.5.

It is immediately noticeable that, despite the differences in distributions, the Type I error rates are in a very narrow range, even for small samples. The only exception is the Yuen test for samples of size ten. Here both the UES.AE and UES.PU have Type I error rates below 0.4. This is due to the very narrow peak in both of these distributions (as seen for AE in Figure 18.4c), which means that for small samples, when the extreme values are trimmed, it tends to give very similar trimmed means that are not significantly different.

It is also worth noting that, again, for small samples, the BM test is a little liberal, being centred around 570 rather than 500 for small samples. And the Cliff test is somewhat conservative for both smaller and larger samples.

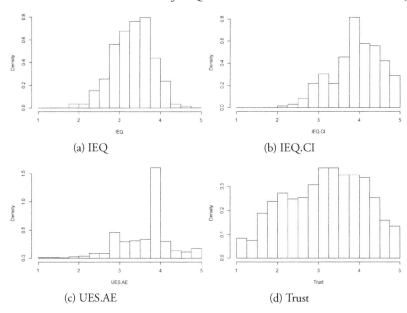

Figure 18.4   The varying density distributions of four of the datasets used in the analysis.

Therefore, for smaller samples and realistic datasets, it would seem that all of the tests are equally good though the Cliff may be a little severe.

18.2.2   Power Analysis

To test for power, a single trial consisted of samples of the same size taken from different distributions. As there are sixteen distributions being considered, this gives rise to 120 distinct pairs from which samples can be taken. Again the samples were compared using each of the six tests and 10,000 trials were run.

It may seem odd to compare samples from completely different scales to see if they differ. However, the goal here is to capture the behaviour of tests when presented with realistic shapes of distributions. The shape is more important than the meaning of the scales. This sort of analysis follows the intent behind work, such as that of Sawilowsky and Blair (1992), but also acknowledges that in realistic situations, both the shape and the location of data can change between conditions. It is already known theoretically that these tests *all* work well for equal sample sizes when the location changes

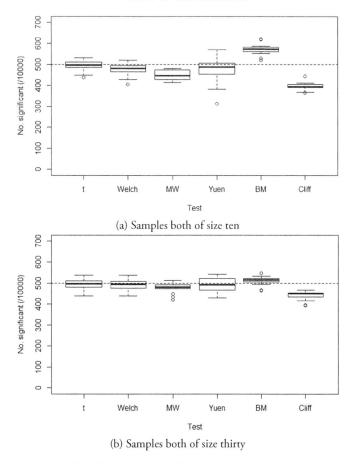

(a) Samples both of size ten

(b) Samples both of size thirty

Figure 18.5   Boxplots of number of significant Type I errors made by each test over 10,000 trials for each questionnaire distribution and for different sample sizes. The dashed line is at 500 tests corresponding to $\alpha = 0.05$.

but the shape of the distribution is otherwise unchanged (see Chapters 11 and 10).

Again to simplify presentation initially, boxplots are used to represent the general distribution of power across the tests, Figure 18.6.

As for Likert items, all six tests give very similar distributions of power with the Yuen test a little lower in power for small samples, a difference which disappears for samples of size thirty. This agrees with the findings of Bakker and Wicherts (2014) in the absence of outliers.

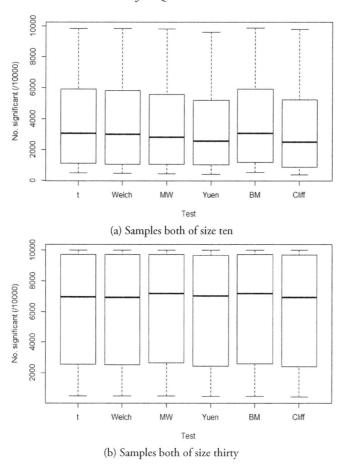

(a) Samples both of size ten

(b) Samples both of size thirty

Figure 18.6   Boxplots of number of significant trials out of 10,000 trials for each pair of distinct questionnaire distributions and for two different sample sizes.

Following the analysis above for Likert items, a change in location was represented as a difference in means for the distributions being compared and dominance of one distribution over the other was also used. As before, the power of the best test out of the six was compared to these ideal measures of difference to see which, if either, was best being captured by the tests. The scatterplots for these comparisons are give in in Figure 18.7.

What is immediately striking about these plots is how closely the value of dominance is a predictor of the power of the best test (though obviously not a linear predictor). This holds even for small samples and across the full

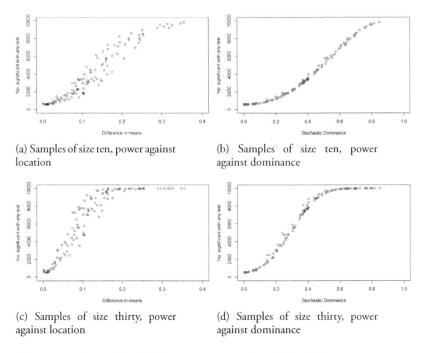

(a) Samples of size ten, power against location

(b) Samples of size ten, power against dominance

(c) Samples of size thirty, power against location

(d) Samples of size thirty, power against dominance

Figure 18.7   Scatterplots of highest power test against the ideal difference in means or dominance between the distributions.

range of dominance values. Though the difference in mean follows broadly the same S-shape, there is much more dispersion: for the same difference in means, the power of tests can vary widely depending on the shape of the distribution. This perhaps reflects more than anything else that if the shapes of the distributions change then interpreting a test result as a change of location is not sound. However, it seems that dominance captures very accurately the meaning of significance in this context.

A very similar picture is in fact seen for each individual test with dominance being strongly related to the power of each test with this distinctive S-shaped curve. The graphs are omitted because they all look essentially like Figure 18.7b. Only the t-test shows a noticeable, but still small, degree of dispersion from the S-shaped curve.

For some of these pairs, the power is very low. These pairs do not necessarily have the same shape, such as the IEQ.CI and the UES.PU, but their means are very similar and one does not really dominate over the other. The result is that none of the tests can easily distinguish between samples taken from each.

In terms of which test gives the highest power, the picture is very similar to that for Likert scales. For small samples, the BM test has the highest power in 62 of the 120 pairs with t-tests giving the highest power in 41 pairs, with no overlap with the BM test. For samples of size thirty, the t-test has highest power on forty-eight pairs and the BM on thirty-four. It should be noted that the Yuen test has highest power on thirty-nine pairs but least power on fifty!

18.2.3   Which Test for Questionnaires?

Just as for Likert scales, there is very little to choose between the tests, with all showing good robustness for Type I error rates, particularly for large samples, and comparable power. This finding matches that of Bakker and Wicherts (2014). They found that MW, BM and Cliff all had comparable behaviour and were slightly more powerful than the Yuen test even when outliers were present. They went on to recommend the Yuen test though, because of the possibility of heteroscedasticity that they did not examine. Such (albeit modest) heteroscedasticity was present in these data but even so the tests behaved comparably. It seems that in the context of questionnaires, there just is not the opportunity for truly unruly behaviour of the underlying distributions and the conservatism of the Yuen test is not needed.

For me though, the real surprise is the very close link between dominance and power. If significance of any test, in the context of questionnaires, is really determined by differences in dominance, then tests for differences in dominance should be used. As such, both the MW, BM and Cliff tests are suitable though noting Bakker and Wicherts' (2014) caution about the MW test and the slight severity of the Cliff test, it seems to me that the BM, Brunner–Munzel test is the test best suited to analysing data arising from questionnaires.

18.3   One Final Observation

Though these simulations give a rather clear result, one further thing arose in my analysis. In the Likert-item analysis, strong multimodality was actually easy to spot in a histogram even with small samples, but it drastically reduced the power of the tests. Of course multimodality makes both location and dominance odd measures because multimodality implies two underlying groups of participants. It may be better to separate out the two groups before trying to see how they differ in the different conditions of the study. Thus, as ever, it is important to look at the data arising from questionnaires with basic plots like histograms. And if you have reason to suspect multimodality, it would be better to work on unearthing why that is happening than to blindly persevere with testing the data.

References

Bakker, Marjan and Jelte M. Wicherts (2014). 'Outlier removal, sum scores, and the inflation of the type I error rate in independent samples t tests: The power of alternatives and recommendations'. *Psychological Methods* 19.3, pp. 409–427.

Carifio, James and Rocco Perla (2008). 'Resolving the 50-year debate around using and misusing Likert scales'. *Medical Education* 42.12, pp. 1150–1152.

Christophersen, Timo and Udo Konradt (2011). 'Reliability, validity, and sensitivity of a single-item measure of online store usability'. *International Journal of Human-Computer Studies* 69.4, pp. 269–280.

De Winter, Joost C. F. and Dimitra Dodou (2010). 'Five-point Likert items: t test versus Mann-Whitney-Wilcoxon'. *Practical Assessment, Research & Evaluation* 15.11, pp. 1–12.

Denisova, Alena, A. Imran Nordin and Paul Cairns (2016). 'The convergence of player experience questionnaires'. *Proceedings of the 2016 Annual Symposium on Computer-Human Interaction in Play*. ACM, pp. 33–37.

Hart, Sandra G. (2006). 'NASA-task load index (NASA-TLX); 20 years later'. In: *Proceedings of the Human Factors and Ergonomics Society Annual Meeting*. Vol. 50. 9. Sage Publications Sage CA, pp. 904–908.

Hudson, M. and Cairns (2014). 'Measuring social presence in team based digital games'. In: *Interacting with Presence*. Ed. G. Riva, J. Waterworth, and D. Murray. de Gruyter, pp. 83–101.

Kamollimsakul, Sorachai, Helen Petrie and Christopher Power (2014). 'Web accessibility for older readers: Effects of font type and font size on skim reading webpages in Thai'. In: *International Conference on Computers for Handicapped Persons*. Springer, pp. 332–339.

Kaptein, Maurits Clemens, Clifford Nass and Panos Markopoulos (2010). 'Powerful and consistent analysis of likert-type rating scales'. *Proceedings of the SIGCHI Conference on Human Factors in Computing Systems*. ACM, pp. 2391–4.

O'Brien, Heather L. and Elaine G. Toms (2010). 'The development and evaluation of a survey to measure user engagement'. *Journal of the American Society for Information Science and Technology* 61.1, pp. 50–69.

Robertson, Judy (2012). 'Likert-type scales, statistical methods, and effect sizes'. *Communications of the ACM* 55.5, pp. 6–7.

Sawilowsky, Shlomo S. and R. Clifford Blair (1992). 'A more realistic look at the robustness and Type II error properties of the t test to departures from population normality'. *Psychological Bulletin* 111.2, pp. 352–360.

Index